D0371735

Effective Safety and Health Training

Barbara M. Hilyer
D. Alan Veasey
Kenneth W. Oldfield
Lisa Craft McCormick

Center for Labor Education and Research
University of Alabama at Birmingham

CRC Press
Taylor & Francis Group
Boca Raton London New York

CRC Press is an imprint of the
Taylor & Francis Group, an informa business

Library of Congress Cataloging-in-Publication Data

Catalog record is available from the Library of Congress.

© 2000 by CRC Press

No claim to original U.S. Government works
International Standard Book Number 1-56670-396-4
Printed in the United States of America   3 4 5 6 7 8 9 0
Printed on acid-free paper

Preface

All four authors of this book are full-time worker trainers who came to this profession from four different directions. All the authors have formal education in the sciences, from chemistry to biology to geology, and graduate degrees in the public health sciences of occupational safety and health, industrial hygiene, and environmental health science. The degrees prove we can read and memorize abundant information long enough for the test, and can connect the dots when required to do so. Several of us also have education degrees of one sort or another, from which faint light occasionally glimmers. From our public health classes, we retained a great deal of scientific knowledge about workplace hazards. Our work experiences in industrial and field settings are varied. We range in age from Generation X to Grandparent.

More importantly to us and to the reader, we have spent a lot of years training all kinds of workers in all sorts of jobs where hazardous conditions are the norm. Our early trainees were most kind, and put up with our dogmatic, fumbling attempts to convey knowledge to people who probably already knew most of it. The people we train have provided us with all the material for this book — good experiences and bad — and it is from them we have learned to do what we do well.

This book is dedicated to the workers who have taught us how to train.

Contents

chapter one

Introduction

Once upon a time, the authors of this book went on a training adventure. We were asked to teach a class at a large pulp and paper mill in a beautiful land along a lovely river. All the workers lived in the nearby small town, and many were third-generation papermakers at that mill. How rural was it? Well, rooms at the only motel went for $17 a night, and one of us shared a wall with a group of duck hunters and their dogs whose river-enhanced furry odors mingled charmingly with the cigar smoke that wafted under the connecting door (Figure 1.1).

Figure 1.1 The training adventure included wet dogs, cigar smoke, and a hostile class.

It was a beautiful place, and the mill was well run and boasted modern, state-of-the-art equipment and an excellent safety record. The workers, however, were not kindly inclined toward the "outside experts" who had been hired by the new safety director (also an outsider) to provide mandatory training in potentially hazardous, non-voluntary job tasks. Several of the enforced participants had just finished a 12-hour overnight shift. We faced an audience of crossed arms, sunglasses hiding shaded eyes, and not a welcoming smile in the room.

To make a long story short (as trainers must always do): By the time our week was up we had trainees stretched out on the carpeted floor to relieve bad backs, invitations to go on fossil-hunting trips, and a lunchtime Pink Floyd Laser Light Show compliments of a class member who had attended the concert and purchased the video. Hostilities faded; experiences were exchanged; and hands were shaken all around on the final day.

We learned a lot that week, and the most important lesson was this: we are all in this training business together. We know a few things about health and safety; workers know a lot of things about the jobs they do; and together, we can do a good job of keeping people safe at work. If we can get workers to tell us what they think they need from us, we will do our best to get it for them. This lesson applies not only to safety training, but to every kind of training or teaching one offers to adults.

I. What you can expect from this book

Most workers look forward to attending a training session about as much as they do going to the dentist, which leave trainers feeling unappreciated and frustrated. This book is written with the goal of turning trainers into Santa Claus, so that everyone looks forward to their visits (Figure 1.2). The presents these trainers bring are interesting topics and welcoming methods that invite trainees' participation in the class. Their gifts will be wrapped in respect for the workers they train, and beribboned with consideration of workers' needs.

Figure 1.2   The trainer can be Santa Claus.

This book is designed to take trainers from wherever they are in experience, attitude, and training responsibilities to a higher place where the training they provide is effective, appreciated, and yes — even fun. The book is structured to take an instructor from ground zero through the process of planning, preparing, delivering, and evaluating an adult training session of from one to one hundred hours. The goal set for trainers will be for their trainees to achieve the basic goals of understanding and retention of the information, enjoyment of the course, and application of the knowledge to their work and their lives.

A. Not your father's training methods

As you were learning all the facts that got you out of school and into the position you now hold, you were the passive recipient, for the most part, of the words of someone who officially knew more than you did. Now you are that official someone, ready to speak the words to a new group of passive recipients.

It won't work. Even though the lecture method is the one with which we are all most familiar, having been exposed to it throughout our formal educational lives, it doesn't work when we are training workers. The premise of this book — that we can train workers effectively while everyone, including the trainer, has fun — precludes lectures except briefly and on rare occasions. There are many, many other ways to train: you can find them all, together with their advantages, disadvantages, and uses, inside these covers.

The most effective methods for training workers are grouped under the title "participatory methods." Just as the name implies, when these methods are chosen workers are invited to actively participate in the learning process (Figure 1.3). They pay attention, remember more, and can more effectively transfer the knowledge to their jobs when they actively participate in the class. They certainly enjoy the training more, and will gladly say so.

In Chapter 2, we build a structure that serves as a framework for training. Chapter 3 describes theories about how workers learn and may help you motivate trainees to change behaviors. In Chapter 4, you hear from workers about their needs and preferences regarding written training materials, and Chapter 5 describes over 30 training methods and ideas. Chapter 6 introduces you to a new, wild bunch you may already have met: "Generation X" trainees, who bring a new set of values and expectations as they invade the workplace and classroom. In Chapter 7, we help you design participatory training activities, and in Chapters 8 through 10, we describe ways to implement and evaluate training. Chapters 11 and 12 are about the people who influence training, and how they all can work together to improve it. Chapter 13 describes activities for training trainers.

B. Sharing control

Sharing control of the class is a little scary at first. Many trainers, especially in the beginning of their training careers, are frightened and uncomfortable in front of a class. It would be unusual not to feel this way. We have all known

Figure 1.3   The most effective training methods are participatory.

this weak-kneed panic, so we seek to control the group by doing all the talking. We don't allow students to ask questions we may not be able to answer. We worry that a trainee will tell a long-winded story, or air unpleasant feelings about management, and we will be unable to handle these interruptions.

These are real concerns, based on situations that have happened to all trainers. In order to prevent this loss of control, many trainers set up a structure in which there is no room for class participation. We agree that someone has to be in charge of a training session, and the logical person to do so is the trainer. We do not agree that the only way to control the group is to fill the allotted time with words that trainees passively receive before they gratefully dart out of the room to go to a more pleasant place.

Another location, another adventure. Barney, recently laid off from his job, returns from a long, liquid lunch to a union-based training session. He sits down at the front of the room facing the class, and starts on a long and loud harangue against the company that fired him. The trainers (two of the authors) are dumbfounded, and have no idea how to regain control from this inebriated, angry, loquacious man.

How did we handle it? Not very well. In Chapter 11 we describe what happened, as well as how we have successfully and unsuccessfully handled many, many other problems (like the classroom fist fight that threatened to break out; the slides that advanced by themselves; and the freight service that lost our materials for an out-of-town class.)

In this book, you will find some ideas you can try in pursuit of group control — methods that allow input from the training participants as they use their minds and experiences to help you find solutions to the problems you are there to address.

In Chapter 12, we pay homage to the other people who frighten trainers — those who judge the efficacy and cost-effectiveness of training, and hold the purse strings to your funding. Although you must retain control of the important things — methods, classroom comfort, training materials — if you will invite certain other people and groups to help you plan and carry out training, you will gain the allies you really need to keep the program vital and ongoing.

C. Outcomes are the focus

The problems trainers encounter, though memorable, are not the focus of this book. The important thing about training is whether trainees learn, and learn to apply, the information and skills they need. To do that they must feel welcome, be comfortable, and be able to access the information you offer. We devote considerable space to describing various means to help people feel comfortable so that they are free to learn, and methods for making information accessible to the many kinds of trainees you will encounter. And you must not lose sight of what satisfies you, the trainer, and what satisfies those who are paying for and making decisions about your training. We will try to deal with all these needs in this book.

D. Application and adaptation

The book includes examples of everything we discuss. Since the book is based on our experience as trainers, so, too, are the examples. All of the worksheets, exercises, evaluations, and outlines are extracted from our own experience. Your training may include other kinds of topics, and you will need to adapt the examples. We have not included anything we have not used and found to work, and so may not have provided you with examples from certain disciplines. You should be able to transfer the guiding principles to your discipline.

We believe the following to be true.

- Every worker deserves a safe and healthful workplace. Safety and health are not inconsistent with productivity.
- Each worker must be provided the skills and understanding to make informed choices about his or her safety and health.
- Training should be provided by people who know the job, understand the hazards and means of protection from these hazards, and are able to facilitate workers' acquisition of knowledge and understanding.
- Effective trainers have respect for workers and value their work and life experiences.
- Training does not have to be boring.

II. Your own training adventure

Open your mind and set out on your own training adventure. Read the book, and then try a few of the things you believe may work in your situation. Everything in this book has worked, or been tried and found not to work, for us. We have provided effective worker safety and health training to Ph.D.-holding engineers and illiterate laborers, sometimes in the same class. We have encouraged the participation of workers who resented management (and, by extension, us) and trainees who like our classes so much they attend every course we teach. Our classes have included unionists and anti-unionists, devils and preachers, shy flowers and rowdy weeds, the pleasantly agreeable and the borderline obnoxious. All of them have been able to contribute to the group effort in some way.

We suggest you begin your adventure with this mantra: No one knows everything, and nobody expects trainers to be an exception. By gathering a group of trainees with similar needs together, you provide a forum for them to argue and agree, cuss and discuss, define problems and hammer out solutions. It is not necessary for the trainer to have the answers to all the questions that arise and, in fact, that would be intimidating. It is necessary only for the trainer to plan ahead by describing potential outcomes, choosing methods that make those outcomes possible, and giving trainees the structured freedom to contribute their knowledge and experience to making the outcomes useful to the group.

This book is filled with nuts-and-bolts suggestions that will guide you on your training adventure. With lots of effort and a little luck, you will find you can train workers effectively while every one, including the trainer, enjoys the experience (Figure 1.4).

Life's a beach . . .
and then you
train

Figure 1.4   Enjoy the adventure.

chapter two

Design and development of training

In 1998, U.S. organizations budgeted $6.07 billion for training. More than 54 million employees received some formal training during 1,713 million hours of training.[1] With this much training being done, shouldn't we spend some time to be sure we get it right?

The people who are tasked with the design, development, and presentation of training probably should have a lifetime's experience in jobs ranging from industrial hygiene to machine operator to pipe fitter in a paper mill, as well as selected bits and pieces of four or five college degree programs. Since none of us is that lucky (or that old!) why don't we instead use a straightforward approach in which we borrow someone else's proven list of steps and follow them.

I. Steps in the design and development of training

It's really not that hard if you have a list. The list described in this chapter has worked well for the authors.

A. Step one – identify training needs

There are number of sources of information within the company that can be consulted to help you identify training needs. They include:

- List of employee job functions and responsibilities
- Job hazard analyses of these job functions
- Accident and injury records
- Federal, state, and local regulations requiring specific training
- Personal conversations with workers, supervisors, and managers
- Safety programs that lead to exposure and accident records being kept or analyzed.

1. Regulatory training requirement overlap

You will probably find that some training requirements overlap and can be combined into one session. For example, identification and hazard analysis training in the Occupational Safety and Health Administration's (OSHA) Hazard Communication and Awareness Level First Responder standards, and the Department of Transportation's (DOT) General Information and Identification requirement for hazardous materials handlers and drivers can, for the most part, be covered by the same information (Table 2.1).

Table 2.1 Comparison of Training Requirements of Three Federal Standards

OSHA First Responder Awareness Level	OSHA Hazard Communication	DOT Hazardous Materials Employee
Understand what hazardous materials are	Know how to use MSDS	
Be able to recognize presence of hazardous materials in emergency	Be able to detect presence or release of hazardous materials	Be able to recognize hazardous materials
Be able to identify hazardous materials if possible	Know what hazard labels identify	Be able to identify hazardous materials
Understand risks associated with hazardous materials in an incident	Know physical and health hazards	Know accident prevention methods and procedures
Understand own role in employers' emergency plan	Know details of hazard communication program	Be familiar with Hazmat regulations
Know site security and control	Know how to protect self	Know self-protection measures
Be able to use *DOT Emergency Response Guidebook*		Know emergency response information
Make appropriate notification		Function-specific training

A worker described the following incident at his industrial facility. A federal DOT inspector arrived to perform a surprise inspection because the facility receives hazardous materials in rail cars for use as raw product. The inspector requested the facility's DOT training records, but there were none. He was given records for annual refresher training performed as part of the company's Hazard Communication Program; he was satisfied the appropriate material had been covered, and issued no citation.

2. Talking to workers

If you do not regularly do the jobs of the workers you will be training, it is foolish to try to train them without first talking to them. There are two reasons for doing this — you can decide which is more important to you. First, they know better than anyone what kinds of problems they are having, what factors are preventing them from working safely or effectively or economically, and what kinds of training will facilitate the achievement of the goal you want to accomplish. Second, the best way to gain cooperation during training is to include trainees in the pretraining discussion. They will participate more willingly in the classes, learn and retain more, and have a more positive opinion of the sessions, which will enhance implementation of the lessons.

B. Step two – determine goal of each training block

Write down the overall goal of each session or topic. In one sentence, what does the trainee need to be able to do at the end of the training? The sentence is a general statement describing the desired outcome.

Sometimes it seems the stated goal of training must have been "Trainees will spend one and a half hours sleeping in a dark room while a television monitor runs, and then they will sign a paper saying they have been trained." Probably a more honest version of this goal is "The company will document the minimum training we think is required to comply with the OSHA regulation." If you are reading this book, you are not that kind of trainer and this smart-aleck comment does not pertain to you, so here's an example of a well-written goal. "Trainees will be able to recognize and identify chemical, radiological, and biological hazards on a hazardous waste site by looking at containers, labels, and other clues."

When you first set out to write goals, it is helpful to brainstorm with the other trainers with whom you will be working, or with other people who are involved in planning or approving the training session. Ask yourselves, "What are we really trying to do here? Why are we doing this training?" If you are training to comply with a particular regulation, you may simply write "The goal of this session is to get trainees to the levels of competency described in 29 CFR 1910.120(q)(ii)." Then you use that standard's list of competencies to write your objectives.

C. Step three – write specific, measurable objectives

Write learning objectives that will lead to the accomplishment of the goal. Write the objectives in terms of what will be *learned*, not what will be *taught*. For example, "to teach workers how to recognize hazardous chemicals" is too vague and is expressed in terms of what the trainer will do, not in terms of what the trainees will do; the latter is preferred.

Here's a better example of objectives to meet the goal stated above. At the end of this session, trainees will be able to:

- Demonstrate they can recognize different types of containers and be able to predict their likely contents

- Identify stressed or damaged containers
- Explain the Department of Transportation (DOT) hazard class system and recognize placards and labels used on vehicles and containers
- Describe the National Fire Protection Association (NFPA) hazard class system and correctly read NFPA labels to identify hazards
- Discuss other clues to the presence of hazards, such as those that can be seen in the vegetation and soil on the site.

One way to ensure your objectives are written with trainee actions in mind is to head up the list of objectives with the phrase "On completion of training, trainees will be able to..." and then list those abilities. Many school teachers start with the phrase "The learner will...," sometimes expressed simply as TLW. It is also helpful to think about how you will measure whether or not the trainee achieves the objective.

Let's take some not-so-good learning objectives and turn them into good ones.

- ☹ Teach workers to know the chemicals they are working with.
- ☺ The learner will be able to demonstrate he or she can recognize container labels on all chemicals in the work area, and use them to identify the contents.

Write the objective so you can measure its achievement. In the second version, the measurement instrument is clear: The trainer points out labels and asks the worker to tell what hazard class they indicate and which chemical contents are identified on the label.

- ☹ The class will learn about the OSHA Confined Space Standard.
- ☺ The learner will be able to state the OSHA criteria for a permit confined space, list the components of a written confined space entry program, fill out a confined space permit (and you would need to write several more if you want to cover the entire standard in the class).

These are specific, not general, and you can measure each one. You get the idea. With practice, it gets easier. Just remember to start with "the learner will be able to..." or simply TLW to remind yourself, and to always be thinking about how to measure whether trainees can do the items you have listed.

1. Outcome-oriented objectives

As you can see, the goals and objectives of worker training should be outcome-oriented. They should be written in terms of the outcome of training, or what the worker can do after training. You should also consider what workers must know going into the training session in order to master the current topic. If the prerequisite skills are lacking, you will have to revise your training plan to include them (Figure 2.1).

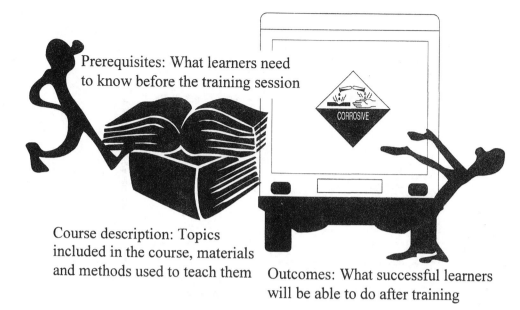

Prerequisites: What learners need
to know before the training session

Course description: Topics
included in the course, materials
and methods used to teach them

Outcomes: What successful learners
will be able to do after training

Figure 2.1 Achieving outcome-based learning objectives.

Writing goals and objectives is probably one of the hardest steps to master for beginning trainers, and for experienced trainers who have neglected this step. It is important to take the time to write them, as everything else flows from these statements. Outlining the course and designing measurement and evaluation instruments follows directly from the objectives.

Remember that each objective must be measurable. You must be able to test whether the trainee can accomplish each objective. A poor objective *might read*, "Trainees will know the company emergency response plan." Since I can't look inside your head to see what you know, I cannot measure this. But I can evaluate what you say, so a better objective states, "Each trainee will be able to state his or her role in the company emergency response plan." Additional objectives would detail other items of knowledge that could be measured by asking, such as the signal for evacuation, the location of a safe place of refuge, and the phone number to dial if a chemical spill is discovered.

2.   How much information?

Your written learning objectives state what the trainee must be able to do to achieve the goal you have set. One of the big decisions to be made is how much information the worker needs to accomplish the objectives. Another way to express this is to say that you may wish to ask the trainees to think at a level higher than rote memorization, and practice so they can make good decisions. This is a good idea when you want to empower workers so they can make informed decisions on their own without relying on orders from

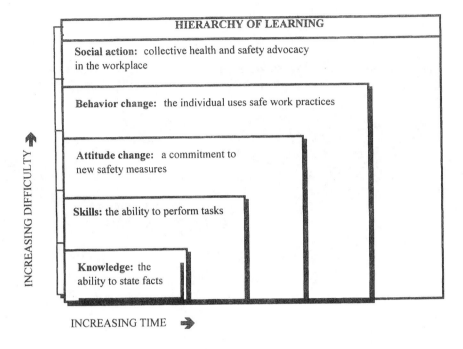

Figure 2.2 Hierarchy of Learning.

others. On the other hand, it takes more time (which costs more money) and sometimes trainees prefer to cut directly to the chase and simply learn what to do.

3. Higher levels of training take more time

At the outset, a trainer should be aware that the higher the level of thinking and using information, the greater the amount of time required to achieve that level. For example, if you want a worker to memorize a list of the nine DOT hazard classes he can probably do it in twenty minutes. If you want him to understand the hazards associated with a spill of chemicals in each DOT hazard class, it will take considerably longer. Which level do your trainees need to achieve to do their jobs safely?

The Hierarchy of Learning developed by educational psychologists helps trainers relate objectives to training hours (Figure 2.2). The boxes represent the relative amounts of training time necessary to achieve objectives relevant to each outcome

4. Writing objectives: the bottom line

What do your trainees need to be able to do to accomplish the goal you have set for them? We return to the "bottom line" of writing objectives. Although one might say a worker cannot have too much training, there are two constraints on offering more information and practice than is necessary to achieve your safety goals.

a. Money. Time is money, and your training budget is not unlimited. Training workers on the clock, as you certainly will be doing, is expensive. Too much time can be boring. If you forget the truism, "The brain can absorb only as much as the butt can stand," your trainees will remind you.

One group of adult learners was asked to rank training factors by how strongly they annoyed the students. Long lectures and periods of interminable sitting ranked high on the irritation scale.

b. Amount of information. Give your trainees only as much information as they need to accomplish the learning objectives you have written. They do not need to know how to do an epidemiology study to understand the application of the data from such studies in their workplace. Adult learners want training with immediate practical application to their lives or their jobs. It may surprise you to know that many people do not enjoy learning just for learning's sake, but want only information they consider useful.

5. Bloom's taxonomy of thinking processes

Bloom's Taxonomy of Thinking Processes provides a clear description of the differences in levels of thinking and the outcomes of learning at each level. When writing outcome-based learning objectives, it is helpful to consider Bloom's Taxonomy.

Table 2.2 shows Bloom's six levels of thinking and defines each level. It differentiates among six groups of outcomes, showing what the student does at the end of training at each level. Ask yourself questions about your trainees. Do you want them to

- Memorize (level one);
- Understand the reasons or relationship (level two);
- Do a simple task or apply what they have learned to several situations (level three);
- Analyze the information (level four);
- Create a new plan (level five); or
- Form new opinions or ideas based on a set of criteria (level six)?

Once you select a taxonomy level that suits your goal, use the verbs in the last column of the table as you write the learning objectives. For most basic worker training activities, a progression to level three will allow trainees to accomplish the stated goals.

The examples in Table 2.3 will help you understand how to apply Bloom's Taxonomy of Thinking Processes to the design of worker training programs. We will assume we want to train the outdoor maintenance crew and there are six different groups of trainees, each with a different goal. The

Table 2.2 Bloom's Taxonomy of Thinking Processes

Level of Taxonomy	Definition	What the Student Does	Verbs to Use in Objectives
Knowledge	Recall or location of specific bits of information	Responds Absorbs Remembers Recognizes	Tell, list, define, name, recall, identify, state, know, remember, repeat, recognize
Comprehension (understanding)	Understanding of communicated material or information	Explains Translates Demonstrates Interprets	Transform, change, restate, describe, explain, review, paraphrase, relate, generalize, summarize, interpret, infer, give main ideas
Application (using)	Use of rules, concepts, principles, and theories in new situations	Solves new problem Demonstrates Uses knowledge Constructs	Apply, practice, employ, use, demonstrate, illustrate, show, report
Analysis (taking apart)	Breaking down information into its parts	Discusses Uncovers Lists Dissects	Analyze distinguish, examine, compare, contrast, survey, investigate, separate, categorize, classify, organize
Synthesis (creating new)	Putting together ideas into a new or unique product or plan	Discusses Generalizes Relates Contrasts	Create, invent, compose, construct, design, modify, imagine, produce, propose, what if
Evaluation (judging)	Judging the value of materials or ideas on the basis of set standards or criteria	Judges Disputes Forms opinions Debates	Judge, decide, select, justify, evaluate, critique, debate, verify, recommend, assess

table states an overall goal for each group. Using the verbs listed in Bloom's table for each level, think of some specific learning objectives you might write for each group.

Table 2.3 Using Bloom's Taxonomy to Write Goals

Level of Thinking	Goal
Knowledge	Trainees can list and locate the parts of a lawnmower
Comprehensive (understanding)	Trainees know how a lawnmower works, and can explain it to other workers
Application (using)	Trainees will teach other workers how to use a lawnmower to cut grass and weeds at a number of plant sites, and repair their lawnmowers if necessary
Analysis	Trainees determine why the lawnmower at the company's main plant is giving the crew so much trouble
Synthesis (creating new)	Trainees build a lawnmower that will work well at all the company's plant sites
Evaluate	Trainees decide what lawnmower(s) the company should purchase

6. The labor empowerment model

Many labor unions with strong health and safety departments design member training programs based on the concept of empowerment for workers. The central tenet of empowerment philosophy is that workers have the right and the ability to participate in the identification, evaluation, and control of workplace hazards.

Table 2.4 shows how the objectives of labor-empowerment health and safety training differ from traditional safety training. If we compare these two sets of objectives in light of Bloom's Taxonomy of Thinking Processes and the Hierarchy of Learning, it is clear that the labor training objectives require a more developed level of thinking, and therefore will consume larger blocks of training time. The first objective under the traditional model can be learned and practiced in 30 minutes, while the same objective under the labor empowerment model would require several days or longer. The choice of objectives, and the verbs used to state them, should be made inside the relative constraints imposed by training time available vs. desirable goals and objectives.

D. Step four – develop materials and methods

Figure out just what information and practice trainees must have in order to accomplish each objective you have written. If you have taken the time to work through writing specific, testable, clear learning objectives, this step will not be difficult. Use the objectives to guide you.

1. Developing materials

To continue with a previous example, look back at the goal and learning objectives stated in step two for the recognition of chemical hazards on a hazardous waste site.

Table 2.4 Traditional vs. Labor Empowerment Model

Traditional Model	Labor Empowerment Model
Workers will be able to identify three steps they must take to make sure their respirators fit properly	Workers will be able to identify a range of control strategies for minimizing exposures to respiratory hazards on the job and to participate in efforts to reduce dust exposures on the job
Workers will learn to follow safe lifting procedures	Workers will learn to evaluate the causes of back injuries, identify control strategies, and advocate for redesign of the work process in order to reduce risks
The crew will be trained to follow the code of safe work practices on the construction site	Workers will develop skills in hazard identification and control in order to participate effectively in the injury and illness prevention program on the construction site

From Wallerstein, N. and Baker, R., "Labor education programs in health and safety" in *Occupational Medicine: State of the Art Reviews*, 9(2). Philadelphia, Hanley & Belfus, Inc., 1994, p. 316. Used with permission.

   a. Goal. Trainees will be able to recognize and identify chemical, radiological, and biological hazards on a hazardous waste site by looking at containers, labels, and other clues.

   b. Learning objectives. On completion of training, trainees will be able to:

- Demonstrate they can recognize different types of containers and be able to predict their likely contents;
- Identify stressed or damaged containers;
- Explain the DOT hazard class system and recognize placards and labels used on vehicles and containers;
- Describe the National Fire Protection Association (NFPA) hazard class system and correctly read NFPA labels to identify hazards;
- Point out other clues to the presence of chemical hazards, such as can be seen in the vegetation and soil on the site.

   In planning for training to meet the objectives for recognizing and identifying hazards on a Superfund site, a team of trainers took the following steps.

   First, the training team reviewed the material in the topics that preceded the hazard recognition chapter. Historical assessment of the site was previously taught, explaining how trainees can narrow down the likely chemicals on the site. Trainees also had learned how a site comes to be listed on the National Priorities List (NPL).

   Second, they wrote fully illustrated materials that covered identifying contents of containers by shape and by configuration of openings, identifying contents by the materials from which the containers were made, and looking at placards and labels on the containers.

Then they produced a videotape of container types, DOT and NFPA labels and placards, and damaged containers. The videotape showed how. a walk-around inspection of the site might indicate, by observation of dead fish or vegetation, where chemicals had been spilled or buried. Many of the slides used in the videotape were of actual NPL sites; the slides were provided by the Environmental Protection Agency (EPA).

Next, the trainers put together a group review packet of actual placards, labels, and various container types so trainees can identify possible contents and hazards as the trainer holds up examples.

Finally, they designed an exercise in which the GHBMO (pronounced "gebmo," the General Hazardous Materials Behavior Model) is used to evaluate and assess the risk of damaged chemical containers.

2. Selecting Training Methods

Development of training also includes selecting the best methods to use for each topic. Consider these factors when you choose teaching methods.

- Content of the material
- Accomplishing the learning objectives
- Achieving variety in the total course agenda
- Progression of hands-on skills
- Possibility of combining several topics in one exercise or practice

Chapter 5 describes training methods in detail, and will help you decide which ones to use.

3. Write a lesson plan

A longtime trainer once said, "Maybe the reason the plan went awry is that there wasn't a plan in the first place." Every topic you teach should have a lesson plan. It is written after the training materials are completed and the methods are chosen. The lesson plan tells you what to do, how long it should take, and what you need to do it with.

a. Why you need a lesson plan. The lesson plan will help you avoid the following embarrassments (all of which happened to us). Of course, the plan doesn't work unless you go over it before and during teaching.

- You get up in front of a class without your overhead transparencies
- You lose track of what you said and teach the same thing twice.
- You forget important words or facts and stand there with a blank face and an open mouth.
- You can't work the projector because you didn't bring the cord. Or the plug adapter. Or a spare bulb, or an extra fuse.

LESSON PLAN: CHAPTER 3, EMERGENCY ACTIONS

Materials: 1993 DOT Emergency Response Guidebooks, 1/trainee
                Overhead transparencies: zoning strategies; incidental vs. Emergency
                Posters: DO and DON'T for Awareness Level (1/local)
                General Industry Standards book provided by the UPIU (1/trainee)
                Erasable red marker

Time: 1.5 hours

There is an important error on page 1 of this chapter. Instruct them specifically to
pick up their pens and change "Part (e)" at the beginning of the second paragraph to Part
(q). Have the trainees look in the standard at 1910.120(q) and highlight the part about the
duties of awareness level responders. Read the competencies as they read along, and
emphasize <u>this is all they should be expected to do</u> in case of an actual or potential
chemical spill. Have them highlight these also on page 1 of the chapter. They have
already learned how to do the first and second of these, and the DOT ERG will help them
do the third.

Take them on a tour through the ERG. Mention there is a newer version but that the 1993
book is quite adequate for their needs. Have them practice individually using the
chemicals listed near the bottom of page 2. When you discuss chlorine, use the overhead
from page 5 to draw in the various zones and distances. Use the overhead from page 6 to
teach hot, warm, and cold zones and what takes place in each. Again reinforce that
awareness level responders stay in the cold zone. Then talk about additional duties of
responders and cleanup crews (page 7) and the training required to do each, using the
overhead. Lead a group discussion about the kinds of actions the trainees have been asked
to do, or are expected to do in their plants. Help them sort out what is and what is not
okay in their examples, and if some of the actions should require more training. Use the
overhead of page 8 to help them understand what an incidental release is: This is
important in industry, and many of our trainees are treating emergencies as incidental
releases. This is especially true of the maintenance people.

Have them work in groups, aided by the experienced member-trainers, and use the ERG
to determine hazards, zones, and actions for the scenarios. These are simple and should
lead to a brief report-back. **These report-backs really drag. If you can find a way to
reduce the drag, please do it.**

End with the poster and point them to page 9, which says the same thing.

Figure 2.3 Lesson plan for trainers who know the topic well.

- You go over your time by an hour and a half because you forgot what the limit was. The class and the next instructor are ready to kill you.

b. Components of a lesson plan. You can decide what you want your lesson plans to include — the things that are important to you. If other trainers will use materials you put together, the lesson plan for these will have to be more detailed than one just for yourself. Ask the other trainers what they want to have included. There are two basic kinds of lesson plans: those for "expert" trainers, and those who are teaching someone else's material and are not extremely familiar with it.

If the trainer writes or compiles his own materials and knows the subject well, he probably does not need extensive notes about how to teach the material. The same is true for lesson plans that will be used by other trainers who are topic experts. For example, the plan might just say, "Describe the OSHA training requirements for Awareness Level First Responders," The trainer already knows what those are. Figure 2.3 shows an example of this kind of plan; it was written for instructors in a small organization where all four trainers were knowledgeable about the material.

Some lesson plans (generally referred to as "instructor's notes") are designed so trainers without expertise can teach the topic. The plan shown in Figure 2.4 is part of a 12-page lesson plan of this type. They are much longer and more detailed, and tell the trainer what to say and do at each step. They are written for lessons with strict sets of visual aids, usually slides or overhead transparencies.

Should trainers who don't know the subject teach using this kind of lesson plan? Many experienced trainers don't think so, even though this type of instruction is used by a number of large organizations. Some trainers call this the "trained chimpanzee" approach, meaning that any primate could teach this material with the lesson plan in hand. Perhaps it could be used by a beginning trainer with subject expertise who wants strong guidance on his first attempt, and could evolve to a style where he is more in control when he is more at ease. Much better training results when the trainer knows the material well and can answer questions and explain situations. Some OSHA standards, such as Hazard Communication and Bloodborne Pathogens, state that expertise and question-answering capability are required of trainers in these topics.

You need a lesson plan. It can include anything you want to include that will help you have all the necessary materials, present your lesson in an orderly manner, and avoid embarrassing omissions and mistakes.

c. Worksheet for a lesson plan. Lesson plans are easy to write. If you are not in the habit, use this worksheet the first few times you write one.

- Name the topic you will teach.
- Write the overall goal of the lesson.

OUTLINE	PAGE/SUGGESTED METHOD

DOT Emergency 12 Response Guidebook	What it is: A quick response guideline for initial use in a chemical emergency. Helpful for about the first 30 minutes
	What it is not: A health effects reference, a firefighting text, a spill cleanup guide, or even the best emergency response document available. It is quick and general, and should keep you out of danger until trained help arrives with equipment.
Pass out the books	Let's use the book to look up the chemical. There are three ways to do this.
Look it up by name.	You have been told the name of the spilled chemical, and it is hydrogen sulfide. Look it up in the blue pages, where chemicals are listed by name in alphabetical order.
	There are two forms of hydrogen sulfide, gas and liquid. Since both lead you to Guide 117, we do not have to know which state ours is in, although we most likely would know in an actual incident.
Look it up by number.	A leaking container has a placard with the number 1053 on it. Look up 1053 in the yellow pages, where chemicals are listed by i.d. number in numerical order.
Look up the label or placard	What if you cannot determine a name or see a 4-digit number? Look up the placard on pages 14-15. The placard for hydrogen sulfide is "Poison Gas," which refers the user to Guide 123 for compressed gases with certain health hazards.

NOTE: This instructor's guide page appears on the left side of the open manual. The appropriate page of the student manual appears opposite it on the right.

Figure 2.4 Lesson plan for beginning instructors, or those without subject mastery.

- Write down the learning objectives: upon completion of this topic, the learner will be able to....
- List the prerequisites of what the learner must know before you start this topic.
- List the method(s) you will use to teach the topic, along with reasons why you chose them. (This will help you justify your choice of methods, and encourage you to use variety in teaching.)
- List the materials you will use.
- Make an educated guess about how long it will take to teach the topic. After you teach it the first time, change the time estimate if you need to.

E. Step five – conduct the training

Consider your options for the location, time, set-up, and all the materials you will need to do the training. You will probably need to locate the following, at a minimum.

1. An indoor classroom

The optimum arrangement includes comfortable chairs at tables (Figure 2.5), since trainees will be doing hands-on work. You will also need audiovisual equipment to present visual aids clearly on a large screen; lighting you can control for the best viewing of visual materials; pleasant temperature; and whatever it takes for your trainees to feel welcome and comfortable.

Comfort is enhanced by having a nearby bathroom, coffee and soft drinks, and a thermostat you can control without having to call maintenance. Comfort level correlates strongly with the informality of the session and the respect you show trainees. There will be more about this in Chapter 11.

2. Space and equipment for hands-on training

This may be an area within the classroom where you stand up a lockout-tagout training board with electrical boxes and connections, or an empty vessel to use for confined space entry practice. It can mean a large field with built-in training devices (Figure 2.6).

3. Audiovisual aids and hands-on workshop devices

These materials are limited only by your resources and imagination. The greater the variety and visual impact of your materials, the greater the impact they will have on your trainees.

If you can invent and create training devices, or have someone on your staff who can, you materials and equipment can be constructed cheaply. The less you spend on each item, the more equipment you can buy, borrow, or build. One trainer who always has fun when she trains created an elaborate "Hazmat Trivial Pursuits" game with color-coded questions in chemistry, health hazards, and other Hazmat topics, using nuts and bolts as the tokens

Figure 2.5 Trainees need tables for learning activities.

on an overhead transparency game board. Teams of trainees rolled fuzzy rearview-mirror dice, and for each correct answer were able to remove one colorful "leak" from a paper drum on a flannel board to stop all their team's leaks and win the game. The game was created BC (Before Computer) and uses construction paper, felt-backed tape, and a flannel board. Modern trainers might do the same thing with presentation software.

One training organization's most popular training aid is the "Leak Monster" (Figure 2.7) that was built from a discarded water heater (instructions for building a Leak Monster are given in Chapter 8. It has six different leaking spots that require six different control measures and a team effort. It sprays water everywhere to add to the excitement, and when all leaks are stopped the team is rewarded by the pouring forth of brightly colored ping pong balls from a pipe on the top. Everybody loves the Leak Monster, and it probably cost under $10 to build.

F. Step six – evaluate and improve training

A training topic can be updated each time you use it by inserting new technology and safety information. Good evaluation instruments will help you determine what materials are being understood and what training methods are working. The people who can give you the most help in evaluating your training are the trainees themselves, other trainers, and other people

Figure 2.6 The training field is designed for hands-on practice.

Figure 2.7 Everybody loves the Leak Monster!

you know who are knowledgeable about the technical aspects of the topic. We will discuss evaluation in Chapter 10.

II. Summary

Starting with a good design and following it through the six steps will make or break your training program. Yes, it is time consuming. No, you don't immediately go into the classroom, so start planning early. The design and development of training will probably take longer than any other part of the training process. If it is quick and easy, you probably aren't giving it enough attention. Use the six steps in this chapter, adapt them to your special needs, and quality training will be the outcome.

In 1998, 70% of the formal worker training in the U.S. took place in classrooms with live instructors. Approximately one third was delivered by outside contractors, and slightly more than one third was designed and developed by contractors. The expenditure for "off the shelf" materials was $2.23 billion, with $2.18 billion spent for custom materials.[1] Contracting for training, or buying packaged training programs, doesn't get the training director off the hook for following the six steps. Start by assessing your needs, setting goals and learning objectives, and then look for contractors or programs that meet the criteria described in these lists. Evaluate before you buy. Be sure your training materials, whoever designs or provides them, measure up to your own standards for quality. This book can help you make determinations of quality for all the training you design, deliver, contract, and purchase.

References

1. "Industry Report 1998." *Training*, 35(10): 47–76. October 1998.
2. Wallerstein, N. and Baker, R., "Labor education programs in health and safety" in *Occupational Medicine: State of the Art Reviews*, 9(2). Philadelphia, Hanley & Belfus, Inc., 1994. With permission.

Interview

I'm Barbara, and I'm a trainer. I used to teach university students — that experience helped me lose my fear of teaching, but it didn't do much to prepare me to be a good trainer. I was pretty serious and boring when I started — I hope I've become more interesting over the 12 years I've been doing this.

I try to establish a commonality with people in the class, because I work hard to promote the idea that we're all in this together to solve a problem by combining all our knowledge. I don't dress up, because the class members don't. Although my degrees may give me credibility when I'm talking about what goes on inside a body following chemical exposure, they might distance me from some people. I'm happy to confess that I hated chemistry until I

found a practical use for it in predicting Hazmat outcomes, and that I almost failed every chemistry course I took.

My personal and political views have no place in safety training. Because of the way I spend my leisure time, I hold a strong pro-environmental position, but I've worked to shut my mouth when a trainee complains about pollution regulations. The environmental argument does not belong in most safety classes, and would serve only to distance me from some trainees and distract the class from the issues we are there to focus on.

I confess I don't spend a lot of last-minute prep time on things I wrote myself or have taught a lot, although I do write bomb-proof materials and gather interesting training aids way ahead. The more action and activities that are associated with any topic, the better. I try hard to read the mood of the class, and if things aren't going well, I immediately change my plans. For example, in a class where the students are not helping me out with the discussion I'll quickly come up with a way to divide up the information, form groups, and ask each group to do an activity that leads to their teaching the material to the rest of the class. Now they are participating.

I watch trainers who are listened to by students, and try to figure out what they are doing. My best role model taught with us for several years. Don is passionate about everything he does; he is emphatic, dramatic, and fascinating to trainees. He moves around the room, waves his arms, yells and whispers — and students are spellbound. He inspired me to be more dramatic, to take a chance and not worry about making an idiot of myself. I try to train with a certain amount of drama: it seems to get people involved.

I love creating new training materials, and I never get tired of training. I grow very fond of people I get to see again and again in classes. Most of the time, I can't believe they pay me to do this.

chapter three

Human behavior and the health and safety trainer

Chances are, if you're old enough to read this, you've heard questions such as "Now why the hell would a guy do something like that?" or "How in the world could anybody do something so stupid?" Usually, these are rhetorical questions. They serve as comments on just how strange the behavior of others can seem.

One of the authors recently spoke with a friend who was having one of those "why the hell" moments. Randy is safety director for a company that takes worker safety and health quite seriously. Recently, while doing air monitoring on an asbestos remediation job, he spotted a worker smoking a cigarette in a contaminated area with his respirator facepiece shoved back on his head (Figure 3.1). According to Randy, the worst of it is that his company recently spent "big bucks" providing respirator training. "For all the good it did, we could just as well have flushed the money down the drain," laments Randy, adding "the really weird thing is that Bill was one of our best trainees. He knew all the answers in class and made an 'A' on the test."

Randy got a painful lesson in the importance of considering human behavior in developing training. What we as trainers try to accomplish centers on the behavior of our trainees. We hope that they will begin safer, healthier practices and abandon unsafe, unhealthy actions as a result of training. To accomplish this type of behavior change, it's not enough just to provide workers with the knowledge required to act safely. Human behavior is complicated by beliefs, attitudes, motives, and values. To develop effective training, we have to answer that ancient question, "What makes people tick?" as applied to the group of workers we need to train.

In this chapter, we will look at some basic theories or models developed by behavioral scientists studying health behaviors. We will also look at how we can use these theories to explain Bill's apparently bizarre behavior and to make training more effective for guys like Bill.

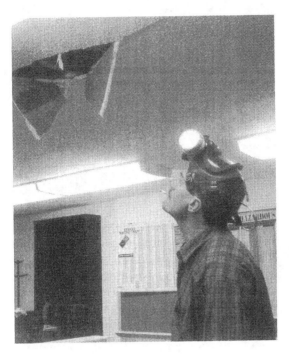

Figure 3.1   Did training lead to safer work practices?

I. Basic considerations for modifying human behavior: stimulus response theory

If you had a psychology course in high school or college, you probably remember Pavlov and his dogs. Pavlov conducted experiments during which he presented food (a stimulus) to dogs in order to study their rate of saliva-tion (an involuntary response). He discovered that if he rang a bell when the food was presented, the dogs would eventually respond to the sound of the bell by salivating, even when no food was present. Based on his research, Pavlov developed what we now know as classical conditioning.[1]

Classical conditioning holds that an unconditioned stimulus (such as presenting food to a dog) produces an unconditioned response (such as the dog salivating), and that a conditioned stimulus (such as ringing a bell), when paired with an unconditioned stimulus, can produce a conditioned response (such as the dog salivating in the absence of food). Based on classical conditioning, researchers have actually been able to control what were previously believed to be involuntary actions in human subjects.

If you recall Pavlov's dogs from introductory psychology class, then you probably also remember B. F. Skinner's operant conditioning. Skinner based his work on classical conditioning, but added to it the idea that behavior is strongly shaped by positive or negative reinforcements that immediately follow the response.[2] Actions (responses) that are frequently followed by a positive reinforcement (reward) have a high probability of being repeated or learned. Responses that are frequently followed by a negative reinforcement (such as punishment) have a low probability of being learned. Classical and operant conditioning fall within what we now refer to as stimulus response (SR) theory. In SR theory, no cognitive concepts such as thinking or reasoning are required to explain human behavior.[3]

Admittedly, stimulus-response theory has a certain common sense appeal, as when rewards and punishments are used to shape behavior. However, SR has limited usefulness in producing meaningful change in workplace safety and health behavior. Some employers claim a high degree of success in promoting safety through the use of rewards based on the number of days worked without accidents, but workers often claim that these programs actually reduce the reporting of accidents and sometimes cause injuries to be concealed.

Another drawback to SR theory is that reinforcements must follow responses immediately and consistently to be effective in changing behavior. In actual workplace situations, the outcomes of workers' behaviors, be they positive or negative, may be neither immediate nor consistent. The lack of immediacy is well represented by chronic exposure hazards, in which decades may elapse between initial exposure and the associated disease. The lack of consistency is apparent in the fact that unsafe acts don't always result in injuries. In some cases, workers may repeat unsafe acts many times without injury resulting.[4]

Think about the negative consequences of Bill's asbestos exposure and cigarette smoking from the standpoint of SR theory. We know that asbestos exposure is associated with very severe negative consequences, such as asbestosis and lung cancer. We know that cigarette smoking is strongly associated with various diseases, including lung cancer. We also know that cigarette smoking and asbestos exposure together have a strongly synergistic association resulting in a significantly greater chance of lung cancer. However, none of these dire consequences would serve as strong negative reinforcements according to SR theory, because they don't appear for many years following exposure. In fact, Bill probably gets immediate positive reinforcement from his behavior. He gets the tight-fitting respirator off his face and relaxes with a cigarette for a few minutes, which he finds very enjoyable. Thus, while SR theory may provide insight into Bill's behavior, it doesn't give us a lot of help in trying to change it.

II. Basics of health behavior models: expectancy value theory

We can begin to address the impasse we have reached in approaching Bill's behavior through SR theory by acknowledging that humans are not just "black boxes" that emit behaviors based solely on positive or negative reinforcements. We are able to think, to reason, and to learn from what we observe in addition to what we experience personally. We all have beliefs, attitudes, motives, and values that affect our behavior relating to safety and health, as well as all other aspects of our lives. For the purposes of this chapter, we will consider health behavior to incorporate all of these factors.

All of the models or theories that we will examine are based in expectancy value theory (also known as value expectancy theory). Expectancy value theories are based on the idea that guys like Bill have the ability to form expectations about the consequences of their behaviors. Thus, in using these theories we are not restricted to stimuli, responses, and immediate reinforcements in trying to explain or change behavior.

Expectancy value theories emphasize (1) the role of an individual's expectation that a specific action will produce a specific outcome, and (2) the degree to which the individual values that outcome.[5] The emphasis here is on cognitive processes, such as thinking and reasoning, rather than immediate reactions to positive or negative reinforcements. Because of this, expectancy value theories are also referred to as cognitive theories.

Expectancy value theories hold that we humans tend to select behaviors that we believe will maximize good outcomes and minimize bad outcomes. Simply stated, we tend to choose behaviors that we expect to produce an outcome that we value. For example, consider the following statement by Jeff, a firefighter, Hazmat team member, and Hazmat trainer from California: "When in doubt, I always wear chemical protective gear because I want to be the oldest surfer in the continental U.S." This comment illustrates the basic expectancy value concept. A long and enjoyable life is something Jeff values and he expects to have a much greater chance of achieving that if he avoids chemical exposure.

III. Examples of health behavior theories

In the following sections of this chapter, we will consider important points of several popular expectancy value-based theories. As we do so, keep in mind that some of these theories are quite complex. A full description and discussion for each of them is beyond the scope of this text. None of the models or theories discussed here is considered universally effective. All have advantages, disadvantages, and limitations. As we go through this chapter, start thinking in terms of how theories such as these might apply in understanding, and possibly even changing, your trainees' behaviors.

A. The health belief model

The health belief model is one of the oldest and most widely used health behavior theories.[6] It was developed in an attempt to explain the failure of many people to participate in disease prevention or detection programs. We can examine the basic idea of the health belief model in Figure 3.2.

As applied to occupational health and safety, the health belief model holds that a worker's perception of threat from a hazard is determined by

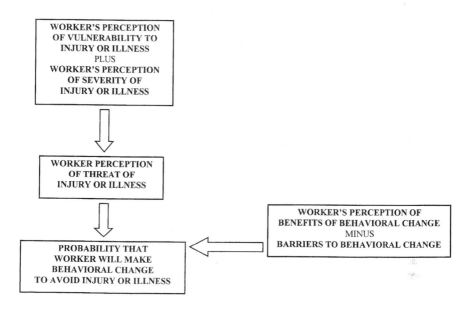

Figure 3.2   The basic Health Belief Model as applicable to worker safety and health behavior.

(1) how serious she perceives the injury or illness that may result, and (2) how susceptible she perceives herself as being to the injury or illness. If she perceives herself as sufficiently threatened, she is likely to take precautions that she believes will prevent injury or illness provided the perceived benefits of the preventive action are greater than the perceived barriers. The basic health belief model has been modified by adding cues to action and modifying factors, as shown in Figure 3.3.

In talking with Bill, Randy learned that Bill doesn't really understand the long term health problems that exposure to asbestos can cause. Bill thinks of respirator use mainly as something required by OSHA. He personally doesn't feel very strongly threatened by exposure, noting that it has never made him "feel bad." Bill also finds the respirator very uncomfortable to work in. Furthermore, he is doubtful about whether or not he is capable of wearing the respirator for a full work period, partly because he can't go that long without a cigarette.

Based on this information, Randy might decide to modify his company's respirator training to include a strong emphasis on the hazards involved, instead of mere instruction on how to use respirators effectively. This part of the training would serve as a cue to action by making Bill aware of his susceptibility to illness and the seriousness of illness that could result from his failure to use the respirator (Figure 3.4). He would develop an accurate perception of the threat posed by the contaminants; this in turn would make him aware of the benefits of respirator use, such as a longer and better life. Also, Randy might try to provide a more comfortable respirator for Bill. This

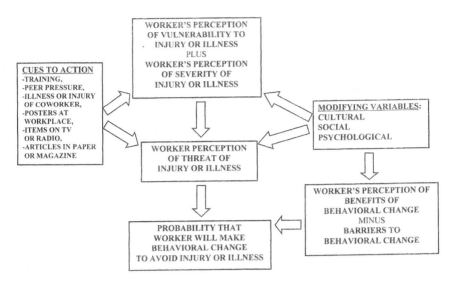

Figure 3.3   The enhanced Health Belief Model as applicable to worker safety and health behavior.

Figure 3.4   The training emphasizes possible health effects of failure to wear a respirator.

would remove discomfort as a perceived barrier to respirator use. If the perceived benefits outweigh the perceived barriers, then the likelihood of Bill using the respirator consistently is high.

B. Social cognitive theory

Social cognitive theory (SCT) acknowledges the role of positive and negative reinforcement, but strongly emphasizes expectations, in shaping behavior. SCT (also known as social learning theory) is therefore strongly rooted in expectancy value theory. SCT holds that whether or not a person performs a given behavior is determined in part by two types of expectations: outcome expectations and efficacy expectations.[7]

Outcome expectations are the beliefs a person holds regarding whether or not a behavior will lead to beneficial outcomes. This is the basic stuff of expectancy value theory, as previously discussed. Efficacy expectations, on the other hand, are the beliefs people hold about their ability to perform a behavior. A person with a high degree of self-efficacy perceives herself as competent to perform a behavior or task.

SCT considers self-efficacy vital if desirable behavior change is to occur. For example, Bill does not currently believe that he is capable of wearing a respirator for an entire work period. That being the case, he almost certainly will not make a serious attempt to do so. According to SCT, in order for Bill's rate of respirator use to improve, his efficacy expectations as well as his outcome expectations must be modified.

According to SCT, self-efficacy can be increased by the following things:

1. Performance accomplishment, through small steps leading toward performance of the desired behavior. For example, Randy might encourage Bill to wear his respirator consistently for shorter intervals that are gradually lengthened into a full work period.
2. Vicarious experience, or learning through observation of others (also known as modeling). For example, Bill sees his co-workers consistently using respirators and decides that he can do so also.
3. Verbal persuasion, for example, words of support and encouragement.
4. Control of emotional arousal, for example, reducing the tension and anxiety that reduce Bill's self-confidence.

C. The theory of reasoned action

The theory of reasoned action (TRA) treats attitude, belief, and behavioral intention as predictors of behavior (Figure 3.5). The TRA holds that the strength of a person's intention to perform a behavior is determined by (1) her attitude toward the behavior, and (2) her perception of what relevant others think she should do (the subjective norm).[8]

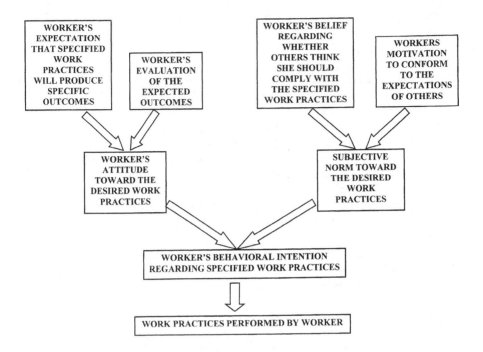

Figure 3.5   The Theory of Reasoned Action as applicable to worker safety and health behavior.

According to this theory, a person's attitude toward a behavior is determined by (1) her belief that the behavior will lead to certain outcomes, and (2) her evaluation of those outcomes. So far, this seems a lot like basic expectancy value theory; however, the role of the subjective norm adds another important dimension. As explained by the TRA, a person's intention to perform a behavior is partly determined by (1) her belief that specific individuals or groups think she should or should not perform the behavior and (2) her desire to comply with their wishes.

Let's consider the problem of Bill's inconsistent respirator use from the standpoint of TRA. In Randy's discussions with Bill, some interesting things emerged about Bill's normative beliefs. To begin with, it was obvious to Randy that Bill is very concerned about what his co-workers think of him. He takes a great deal of pride in feeling that he is respected as a good worker and team player. It was also obvious that Bill believes his co-workers don't have a strong opinion on whether or not he should wear a respirator. This belief is based largely on his observation that others in the workplace don't wear respirators consistently either. In response, Randy might attempt to employ peer pressure in order to increase compliance with requirements for respirator use. By depicting consistent respirator usage as "smart," "a group effort," or "the in thing," Randy may be able to change the subjective norm within the workplace to one that is highly favorable toward respirator usage.

However, if workers such as Bill don't really care what others think about them, this would probably be a fruitless approach.

D. The transtheoreical model

If you think back over the major behavioral changes in your life, you will probably note that those changes didn't occur instantaneously. Such changes often take place in increments or stages that occur over a long period of time. This concept is a major aspect of the transtheoretical model (TTM), which is also known as "stages of change."[9] According to TTM, changes in health behaviors can involve up to six distinct stages, as shown in Table 3.1. Keep in mind that TTM is not a linear model, but rather a circular or spiral one, so we can enter or exit at any stage in the process. Also keep in mind that relapse, should it occur, is treated as just another stage, rather than the end of the attempt to change behavior. You will learn more about this when we get to the relapse prevention model. TTM has been used primarily in changing addictive behaviors. It could be useful in addressing Bill's tobacco addiction, which is one factor in his failure to use the respirator consistently.

Table 3.1 Stages of the Transtheoretical Model

Stage	Description
Precontemplation	No serious consideration is given toward changing behavior in the foreseeable future.
Contemplation	A behavior is seen as a problem and serious thought is given to changing it, but no definite commitment to do so is made.
Preparation	A definite decision to attempt to change the behavior is made and some preparatory actions may be taken.
Action	The actual behavior change is undertaken.
Maintenance	Efforts are undertaken to prevent relapse and to consolidate gains achieved during the action stage.
Relapse	Failure to maintain the behavior change occurs and the undesirable behavior is resumed.

Randy's conversations with Bill indicate that he has never really considered giving up cigarettes and doesn't see smoking as a problem. Based on this, we can assume that Bill is in the precontemplation phase. Training in the hazards of asbestos exposure will provide a good opportunity to raise Bill's awareness about the hazards of smoking, given the synergistic effects of the two exposures. Raising his awareness may very well nudge Bill into the contemplation phase. Tactics designed to emphasize the advantages of quitting and the disadvantages of not doing so might help Bill transition into the preparation stage. One example of such a tactic might be placing posters and pamphlets in the break room that point out the ill effects of second-hand smoke on Bill's family and co-workers (Figure 3.6). In the preparation phase, Bill might reduce the number of cigarettes smoked per

Figure 3.6   Bill reads health and safety pamphlets in the break room.

day or delay smoking the first cigarette of the day for some prescribed time. Upon actually giving up cigarettes, Bill will enter the action phase. In this phase medications designed to ease the craving for nicotine may be helpful. After Bill has been in the action stage for some time, he will transition into the maintenance phase. Keep in mind that maintenance is a continuation of change, rather than the absence of change. Avoidance of relapse is a major goal of this stage, and will be discussed separately in the following section.

The transtheoretical model can be helpful in deciding when and how to try to bring about changes in health behaviors. It can prevent mistakes, such as trying to convince Bill to wear a nicotine patch to stop smoking while he is still in the precontemplation phase.

E. Relapse prevention theory

Relapse prevention theory is actually an aspect of a number of health behavior models rather than a single discernable theory.[10] In some health behavior programs, relapse is considered the rule rather than the exception. After making a behavior change, it is not unusual for a person to relapse, or resume the previous behavior. Multiple relapse episodes may occur before a behavior change becomes permanent. This is why the transtheoretical model treats relapse as a part of behavior change; a regression to a previous stage of change rather than abandonment of the entire process of change.

Relapse can produce a sense of failure, resulting in total abandonment of the program. Relapse prevention theory provides coping skills intended to lessen the frequency of relapse and to prevent it from derailing all attempts at behavior change should it occur.

IV. What does all this mean to the safety and health trainer?

We began this chapter with a description of Randy's dilemma regarding Bill's poor habits of respirator usage. This forced Randy to realize that Bill wasn't adequately trained despite the significant sum that Randy's company spent on respirator training. The training provided to Bill was excellent in terms of the technical aspects of respirator use; however, it failed to bring about the desired change in Bill's behavior. We could say the training ultimately failed, leaving Randy a very unhappy safety director.

The problem with Randy's approach to training was that it focused on the training instead of the trainee. This "one type fits all" approach assumes that the trainee is an empty vessel to be filled with knowledge that will produce instant enlightenment, causing him to perform the desired behavior. The problem with this approach is that none of us is an empty vessel or a robot to be programmed. We all have attitudes, beliefs, motivations, and emotions that determine our behaviors. These basic factors of health behavior are at least as important as OSHA's training requirements as considerations in training development.

In this chapter, we reviewed some of the relevant features of a few models or theories of health behavior. This was in no way intended to be a comprehensive coverage of this topic. Rather, this chapter was intended to provide examples of basic considerations of health behavior that may allow us to understand, and perhaps even change, worker health behavior. Incorporating these considerations into trainee needs assessment and training development makes it much more likely that we can bring about safer worker behaviors through the training we provide.

References

1. Davids, A. and Engen, T., *Introductory Psychology*, Random House, New York, 1975, Chap. 2.
2. McKensie, J.F. and Jurs, J.L., *Planning, Implementing, and Evaluating Health Promotion Programs*, Macmillan Publishing Company, New York, 1993, Chap.6.
3. Rosenstock, I.M., The Health Belief Model: Explaining Health Behavior Through Expectancies, in *Health Behavior and Health Education*, Glanz, K., Lewis, F.M., and Rimer, B.K., Eds., Jossey-Bass Publishers, San Francisco, 1990, Chap. 3.
4. Brauer, R.L., *Safety and Health for Engineers*, Van Nostrand Reinhold, New York, 1990, Chap. 3.

5. Carter, W.B., Health behavior as a Rational Process: Theory of Reasoned Action and Multiattribute Utility Theory, in *Health Behavior and Health Education*, Glanz, K., Lewis, F.M., and Rimer, B.K., Eds., Jossey-Bass Publishers, San Francisco, 1990, Chap. 4.
6. Rosenstock, I.M., Historical origins of the health belief model, *Health Education Monographs*, 2, 328-335, 1975.
7. Bandura, A., Social Cognitive theory and Exercise of Control over HIV Infection, in *Preventing AIDS: Theories and Methods of Behavioral Interventions*, DiClemente, R.J., Peterson, J.L., and Rimer, B.K., Eds., Plenum Publishing, New York, 1994, 25-39.
8. Becker, M.H., Theoretical Models of Adherence and Strategies for Improving Adherence, in *Handbook of Health Behavior Change*, Shumaker, S.A., Schron, E.B, and Ockene, J.K., Eds., Springer Publishing, New York, 1990, Chap.1.
9. Prochaska, J.O., DiClemente, C.O., and Norcross, J.C., In search of how people change, *American Psychologist*, 47, 328, 1992.
10. Marlatt, G.A. and George, W.H., Relapse Prevention and the Maintenance of Optimal Health, in *Handbook of Health Behavior Change*, Shumaker, S.A., Schron, E.B, and Ockene, J.K., Eds., Springer Publishing, New York, 1990, Chap.2.

Interview

I'm Elizabeth, and I'm a program director. I direct a NIOSH center where we hire instructors to teach health and safety courses. Instructors may come to me with an idea, or are referred to me. We pay on a scale; for people who have to develop the lecture and teach, we pay $100/hour. For something like demonstrating how to use a machine, $35/hour. Instructors who teach a whole day are paid around $700/day.

I look for instructors who have enthusiasm about the topic — it is so important for us that whoever's coming in believes in what they're talking about. Most folks that go into the field of occupational health and safety are truly dedicated. Credentialed people aren't always the best instructors, they may have the knowledge but not the training skills. We always evaluate our programs, and if students don't like what the instructor does — specifically, in content and presentation skills — we won't ask that person back. The folks you want to continue to work with are those who ask how they can improve themselves. Doing this for 11 years, I've seen them develop into better trainers because they have a dedication.

When we plan classes we arrange the hotel, the meeting room, everything, away from our center. We need to be aware of what our people want so they'll come. Different trainees like different things; nurses, for example, like convenient parking and clean, safe facilities. In a big city, they are concerned about their safety, so we provide good directions to everywhere. Nurses, too, are outspoken about letting you know if you're not doing what you need to do for them. Some of the other safety folks, like industrial hygienists, don't care — they'll go anywhere. Physicians like to go to the beach, say in a resort area. Professional points are an issue for many of our groups; not just CEUs, but also professional points for nurses, physicians, industrial hygienists.

We try to have a mix of healthy and not-so-healthy food. Good stuff (fruit, juice) and good, fattening Danish, too. We do good afternoon snacks, too — chocolate, coffee. We like to choose a place where people can walk to get a nice lunch. Many people who come to training courses are on a budget, and don't have a car. We need to be able to tell them the hotel will pick them up at the airport. We try to have good lighting in the meeting rooms, because so many instructors now use technical training aids like projector systems.

It seems to me that some people may be getting too technical, they think everything needs to be computer-based. Maybe it's good to be innovative in presentations, but I hope instructors won't forget some of the simple stuff too. Like eye contact; remember these are people you are talking to. Don't just get totally into your remote control - every now and then, put it down.

chapter four

Training materials

The choice of the materials you use in training and the methods you select are extremely important, as they set the tone for your class. These choices determine whether your trainees enjoy the course, and learn what you have stated they must do in order to work better or more safely.

Materials for training workers can include many different items to strengthen the message and make it memorable. These are just a few.

- Videotapes
- Computers, CDs, and disks
- Placards and labels
- CPR and rescue dummies
- Wrenches and other tools
- Ropes, ratchets, and tie-downs
- Respirators and chemical protective clothing
- Written information

The trainer's job is to select materials that are useful, interesting, and varied. We will discuss videotapes, written materials, and the general category of classroom props. Just as they do in choosing training methods (discussed in Chapter 5), effective trainers select materials that enchance the topic, suit the training population, and add variety, hands-on experiences, and visual interest to the class.

I. Videotapes

If you have been a trainer for longer than three days, you have been on the receiving end of strong marketing activities by training tape companies. You are aware of the enormous number and variety of tapes available, and have a general idea of their topics and costs. You may have entertained the idea, as most trainers do at one time or another, that you could make your own training tape. Let's talk about cost, appropriateness, accuracy, and attitude of videotapes you may consider using in your classes.

A. Costs

Commercial videotapes range in cost from about $200 to around $1000. To produce a videotape of your own is quite expensive if you want professional quality — one training organization recently was quoted a price of more than $1000 for a five-minute videotape. Professional, high-resolution filming and post-production costs are high. (Of course, brief segments of homemade tapes are quite valuable in illustrating how to do, and not to do, job tasks.) Training tape suppliers frequently have special sales. If you are on their mailing lists, you will receive notices. Most have a reasonably priced preview policy.

Always preview a video. As you watch it, keep in mind the learning objectives you have written, and reject anything that doesn't help your trainees achieve them. View them with certain criteria in mind. Keep notes. If you buy the tape, store your notes inside the cover where they can be helpful to the other trainers in your organization, and serve as reminders to you.

B. Appropriateness

Having trained a number of settings for a variety of programs, we have viewed a lot of training tapes borrowed, copied, and otherwise, obtained for free from various sources. These are often selected just because they were free; sometimes they fit the topic and the group, and sometimes they do not. For example, trainers of hazardous waste site workers have special needs for tapes that show site hazards; these are limited, so they might use chemical spill tapes from industrial settings. The material matches up about 50%, so who worries about the other half? In a case like this, only the trainer can decide. If the tape helps you achieve the learning objectives you have written for that topic, use it. If not, find another way to get the material across. In any case, be sure to summarize the information and relate it to your message.

One good way to attract trainees' attention to a videotape is to give them a quick questionnaire before you show it (described in Chapter 5). Answers to the five-minute questionnaire are found in the tape. Of course, the trainer explains this before running the tape. To be appropriate, tapes must not put trainees to sleep. They should show action, not words. They should be as short as possible. There are differing opinions on length, but never to ask trainees to sit for more than 20 minutes. That includes the introduction to the tape, the showing of the tape, and any words you want to say after it is over.

There are other ways to put people to sleep with videotapes. Show them after lunch. Turn the lights out. Show the tape on a tiny monitor. Provide no introduction to direct trainees' attention to the tape, and no instruction as to how they will be expected to use what they see on the screen.

In one informal survey of several classes, trainers asked for the one worst sin encountered in training. The lengthy videotape won by a landslide, especially when it was the only method in the session, and was followed by a form to be signed by the viewers documenting that they now were trained. One worker reported sleeping through five hours of consecutive training tapes.

C. Accuracy

Unfortunately, not all videotapes are accurate. We have rejected several for containing technical errors; some of these were produced by organizations that should have known better. If you really like a tape for all its other components, you may decide to show it but to explain to the class that you take issue with a certain point. Otherwise, send it back. You are not in business to teach errors in the life-and-death world of safety.

D. Attitude

Surprisingly, some training tapes show bad attitudes. The two bad attitudes most frequently noticed are a lack of reality (we call this "One whiff of this stuff'll kill you" after a trainer who overused this phrase when teaching about chemical hazards) and a patronizing tone. The first lends a nonprofessional air to the tape, and results in a dilution of its warning power as it continues to cry "Wolf!" The second gives one the impression the producers, and especially the actor/narrator, consider workers to be ignorant. Send these back, too, with an explanatory comment.

E. Reality

Many commercial training tapes, however, are quite good if used in small doses, not consecutively, backed up with the appropriate instructor input, and never just after lunch. They show actual conditions, outcomes, and consequences of unsafe behavior, and allow trainees to see how to do an act safely. The really good ones are also entertaining.

In the fire/Hazmat community, there exists a network of people who accumulate very good training tapes from raw news footage of emergency situations. These include fires, explosions, chemical incidents, confined space facilities and rescues, and other aspects of fire and rescue service activities. Within a few days after any local event, the television news crews retain footage of the incident. In many cities, the local stations will give you these if you explain you want to use them for training. After the first few days, the tape may be reused and no longer available. Some of these tapes are horrific, but still can be used if the victims are treated with respect in your classroom. Our library includes footage of a wrecked truck being uprighted with air bags; a huge warehouse fire from which pesticide runoff contaminated many miles of creek and lake; a group of firefighters blowing up a toluene tank with a circular saw cut; a burning house that was built of old batteries (how's that for a products-of-combustion lesson?); and the cylinder from a burning propane truck rocketing down five miles of interstate highway, burning everything in its path.

II. Classroom props

You need props in the classroom. How much better it is to show items of chemical protective clothing than just to talk about them and show pictures. How much easier to learn to recognize DOT hazard class placards when the actual placards are hanging on the walls. In Chapter 8, we will discuss tools and equipment for doing hands-on training, but props are also necessary for presentation training.

Props can be expensive. A label company will sell you a set of placards for under $100; but if you want a totally encapsulating chemical protective suit, you will spend at least $800, and that's for the single-use variety. Air monitoring equipment is pricey, but how else can your trainees learn to use it? If your budget just won't stretch far enough, make friends with local vendors. If your company purchases their equipment, they will certainly provide it for in-house classes; if you do open-enrollment training, they may loan you equipment on the grounds that you are putting their product before the public. You, of course, will state you are not recommending one brand over another (unless you are), and explain that you are happy to receive loans from all vendors. For the helpful vendor, familiarity may breed sales.

III. Written materials

Not all courses require a written training manual. The average worker, if such a person exists, will not go home and read the manual after a class. Don't give trainees more information than they need. There are two reasons to provide written information.

- People do not have to write down the important things you say. With a yellow highlighter, they can spot the parts they want to remember.
- Nobody will remember everything later. Trainees can use the manual to look things up when they need them.

Consider the following characteristics when you choose or produce written materials.

- **Content** — What do they say? What do they teach?
- **Suitability** — Are they easy to read, and in the language of the workers?
- **Interest** — Do they apply to workers' jobs or lives? Are they interesting to look at and to use?

A. Content

Content follows learning objectives. If one of the objectives is for the learners to recognize container labels and identify the hazards, the materials will include labels and the material safety data sheets (MSDS) listing the hazards.

Materials should be as specific in content as possible. Use labels and MSDS your own trainees will work with. Use cases from your facility as examples to make a point. Whether you write or make your own materials, or purchase them ready-made, the same factors must be considered.

B. *Suitability*

Choose materials that fit the trainees in the jobs they do. You may have to adapt the materials to jobs in different parts of the plant, but you should take the time to do this. Use examples from trainees' own jobs as teaching points.

Obviously, if your trainees cannot speak and read English, all your materials must be in their language. The need for training in a variety of languages is growing in the U.S. and is especially difficult to satisfy unless the materials you seek are in Spanish. Because the history of immigration from Spanish-speaking countries is long, there are quite a few sources of Spanish-language materials. Start by calling the AFL-CIO in Washington, D.C. The Labor Occupational Health Program at the University of California at Berkeley produces some Spanish-language materials. The National Clearinghouse for Worker Safety and Health Training in Bethesda, Maryland can provide a list of sources.

Other national and regional languages are finding their way into the North American workforce; materials in these languages are difficult, and often impossible, to come by. As a last resort, hire a translator. If possible, group trainees and workers so they can help each other, including an English reader in each group.

Literacy is a big concern for trainers, and impossible to assess simply by glancing over the class. Statistics validate this concern and, in fact, the situation is probably more serious than you realize.

1. *Literacy in America*

A 1992 survey by the U.S. Department of Education's National Center for Education Statistics estimated that about 21% of the adult population — more than 400 million Americans over the age of 16 — had only rudimentary reading and writing skills. Roughly 4% of the total adult population, or about 8 million people, was unable to perform even the simplest literacy tasks.[1]

In 1994, compared with most of the countries assessed, the U.S. had a greater concentration of adults at the lowest levels of literacy across the prose, document, and quantitative literacy domains. Only in Poland did a greater proportion of the population score at the lowest level across all three literacy domains.[2]

a. Definitions of Literacy. In the National Literacy Act of 1991, the U.S. Congress defined literacy as "an individual's ability to read, write, and speak in English and compute and solve problems at levels of proficiency necessary to function on the job and in society, to achieve one's goals, and to develop one's knowledge and potential."[3] National and international literacy surveys measure literacy along three dimensions — prose literacy, document literacy, and quantitative literacy — designed to capture a whole set of information-processing skills and strategies that adults use to accomplish a diverse range of literacy tasks.

   b. Three dimensions of literacy. Literacy with prose implies that people can locate information, find all the information, integrate information from various parts of a passage of text, and write new information related to the text. Document literacy includes using tables, data forms, lists, graphs, charts, and maps. Quantitative literacy requires that individuals be able to locate and use quantities that may be displayed visually in graphs or charts, or shown in numerical form using whole numbers, fractions, decimals, percentages, or time units appearing in both prose and document forms.[4] All three dimensions of literacy have a place in training manuals, so a consideration of their accessibility is important.

   c. Literacy survey results. The literacy proficiencies of young adults assessed in 1992 were somewhat lower, on average, than the proficiencies of young adults who participated in a 1985 literacy survey. The 1997 survey results were lower still. While in many other countries young adults had higher literacy levels than older adults, distribution of literacy proficiency across different age groups was fairly uniform in the U.S., except for adults aged 56–65, who scored slightly below other groups.[5]
   Table 4.1 shows the results of the 1997 literacy survey for U.S. participants. For explanations of the tasks required at each level, consult the U.S. Department of Education at their web site, www.ed.gov. Keep in mind that in level one on the prose scale, the reader is required only to locate a single piece of information identical to or synonymous with the information given in the question, when the text is short; or when plausible but incorrect information is either not present, or is present but located away from the correct information. In short, success at level one requires minimal reading skills. Of the 21% of survey subjects who were assessed at level one: 4% could not perform even the simplest literacy task. In other words, the eight million people they represent cannot read all.[6]

Table 4.1 Results of Literacy Survey for U.S. Participants: Percentage Distribution of the Population Across Literacy Levels: 1994

	Level 1	Level 2	Level 3	Level 4/5
Prose Scale	20.7	25.9	32.4	21.1
Document Scale	23.7	25.9	31.4	19.0
Quantitative Scale	21.0	25.3	31.13	22.5

   d. Implications for workplace success. According to the National Center on Education and the Economy, "The association between skills and opportunity for individual Americans is powerful and growing. Individuals with poor skills do not have much to bargain with; they are condemned to low earnings and limited choices."[7] But some adults who displayed limited skills in the literacy survey reported working in professional or managerial jobs, earning high wages, and participating in various aspects of society (Table 4.2). The authors have met workers who are successful at supervisory positions, and many who were promoted into these positions because of their successes, who cannot read but have highly developed competencies in accomplishing their jobs without reading. Although privately some admit to getting a lot of assistance from others in performing everyday literacy tasks, the majority of those who demonstrated limited skills in the survey stated they receive no assistance from others.

Table 4.2 Results of Literacy Survey for U.S. Participants: Percentage Distribution on Prose Scale by Profession: 1994

	Level 1	Level 2	Level 3	Level 4/5
Manager/professional	3.9	15.6	37.0	13.0
Technician	3.5	29.9	52.6	14.0
Clerk	6.3	38.0	40.0	15.3
Sales/service	15.9	44.3	34.7	5.0
Skilled crafts worker	29.4	38.0	25.2	7.1
Machine operator/assembler	28.9	36.9	27.8	6.3
Agriculture/primary	31.7	21.2	24.5	22.7

A friend describes his father, who was successfully promoted in a local industry that reconditioned airplanes. "Although my dad couldn't read, he did such good work at his job that he was promoted several times. He became a supervisor. When he needed to read something, he brought it home and my mom read it to him. He even ordered chemicals this way. As he rose up the latter, the company wanted to send him away for some training. He knew that would lead to his being found out, and he kept putting it off for one reason or another. Finally, he retired so he wouldn't have to go. The company never knew couldn't read."

e. Wisdom for trainers. The bottom line is that some workers can't read at all, and many do not read well. Quite a few can't interpret graphs, tables, and charts. Almost none of these poor readers will tell you so. The majority of low achievers in the literacy survey described themselves as reading or writing English well. Your best bet is to assume you will be training people who get their information from some source other than the printed page. You are reminded throughout this book that participatory methods create the best setting for learning. They are the only effective training methods for low-literacy trainees.

Unless you know you will be training a group with extensive formal education and excellent literacy, use the guidelines that follow for creating written materials. Even the good readers will appreciate accessible materials. Assume literacy problems in every training session. Please do not ask, as one trainer was observed to do at the beginning of class, "Hey, any of you guys in here can't read?"

2. Guidelines for written materials

The guidelines that follow are based on two major sources, as well as our own experience. *The Right to Understand* is the best resource we have seen on this subject, and we suggest you get a copy (see references at the end of the chapter). Also, the Agency for Toxic Substance and Disease Research (ATSDR) has developed a style guide for patient education, and provided literacy statistics and anecdotal evidence of how poor literacy leads to confusion on the part of patients.[8] Our own experience in training a variety of workers has taught us that our academic backgrounds in no way prepared us to train, and our early materials were too long, too dense, and too wordy. As always, hands-on failures and successes resulted in knowledge gain and retention. Getting blasted or praised face-to-face or on a student evaluation is good motivation to learn to train better. We apologize and send thanks to all the workers who taught us how to train.

a. Don't write everything you know. "Be complete, but include only the necessary information," say the ATSDR guidelines. Great advice, but a little short on implementation strategy. It sounds so simple, but obviously it is not because the writer is left to define "necessary." What you may consider necessary, another trainer may not. Certainly, the worker may not. This is better settled in training staff meetings to even out the input from all the trainers who will be teaching the material. One solution that works when you anticipate a mixed literacy class is to include the information absolutely necessary for the worker to do his job in the general text, and place extra or interesting technical or background information inside a box. The box has a frame or a pale fill color to set it apart, and trainees are reminded from time to time that the boxed information is not required reading.

Deciding what to put in and what to leave out is one of the trainer's most difficult chores. If you are in this for the long haul, include questions on your course evaluations (see Chapter 10) that will help you with this problem. This assumes your manual is in loose-leaf format so it can be revised. We recommend loose-leaf format for several reasons, the most important of which is ease of revision.

Establish your priority message, then build the text around it. An effective priority message is brief and memorable. In safety training, the message also gives suggestions for avoiding the hazard. Skip the justification for considering the practice hazardous; just tell the worker what it will do to her and how to avoid the bad outcome.

b. Make the page inviting. Break your information into small chunks. Spread it out, leaving plenty of white space. Use lots of heads and subheads, and arrange lists with bullets instead of running them into sentences. Use at least 12-point type; 14 is better if you have the room. Don't type in all caps, and use bold typeface sparingly, if at all. Too many attempts at emphasis are overwhelming, and become irritating. One recently examined publication used bold, italic, and all caps for emphasis, all on the same page, leading one reviewer to toss the book across the room. Don't justify the right margin; let it remain ragged. Right justifying spreads the type and makes it more difficult to read.

Include graphic elements wherever possible. We always sprinkle in a lot of gratuitous clip art — relevant to the text in some way, but not usually to illustrate a point — simply because it makes the page look more interesting. Clip art copyrights are not violated if you use the training manual for your own classes. Invest in a digital camera or a scanner so you can insert photos of your own facility and your own people. Try to have some sort of graphic element on every page. Remember, though, that low literacy trainees will not use graphs and charts; the graphics referred to here are line drawings or photographs.

c. Be brief. Break long sentences into shorter ones. Then shorten your sentences even more by cutting out unnecessary words, phrases, and even ideas. Whenever possible, choose the shorter of two words. If it is necessary to use technical terms, as it often is in safety and health training, define them in parentheses within the sentence.

d. Use active voice, conversational tone. If you are a product of graduate school, especially in a science, it is engraved indelibly on your brain that you must never use active voice. A writer is not allowed to state "We cleared the slides with 70% isopropanol" but must write "The slides were cleared with 70% isopropanol." Get over this. Instead of writing in a training manual that "A Level C protective ensemble should be used," force yourself to tell the worker to "Wear Level C in this situation." It's okay in training materials to write as if you are speaking to the reader, especially when the reader is likely to be a poor reader. You may use contractions, slang, and incomplete sentences if they facilitate the delivery of your message.

e. Illustrations. Workers who were self-identified as poorly literate were asked in focus groups to choose their preferences from several kinds of illustrations. They preferred clean line drawings without a lot of background clutter; larger body areas that include arms, rather than just a drawing of the arm; and illustrations of real objects rather than cartoon figures (for example, a drum with little arms and a scowling face was rated low).[9]

f. Scoring your materials. The following checklist was developed by the authors of *The Right to Understand,** and is used with their permission. It can help you evaluate the writing, design, and illustration of your materials.

Begin by choosing one page of material, one with an illustration. As you answer the following questions, fill in each blank with one of these numbers: yes = 2, somewhat = 1, no = 0.

Verbs

1. Do most of your sentences have active verbs?
   Active: *She did it.*
   Passive: *It was done by her.*
2. Do your sentences have many verbs that are commands, or in the present tense?
   Command: *Write your sentences in the present tense.*
   No command: *You should write your sentences in the present tense.*
3. Are most of your sentences stated positively?
   Positively: *Do it this way.*
   Negatively: *It is not a good idea to do it that way.*
4. Are most of your sentences simple, with no subordinate clauses?
   Example of a sentence with a subordinate clause:
   Carbamate pesticides, which are used on grapes, have been banned in many cases.
   Example of a sentence with no subordinate clause:
   Carbamates are pesticides used to spray grapes. Some are now banned.
5. Are most of your sentences short — fewer than 20 words long?

Meaning

6. Can you identify the priority message?
7. Do you explain any technical words you use?
8. Do you use simpler words where possible?

Design

9. Do you use a large, serif typeface for the main text?
   (Serif type has "feet," as the type used in this book. Nonserif looks like this — no feet.)
10. Do your sentences avoid large sections of capital letters, bold type, or italics?
11. Are the margins wide enough to allow plenty of white space?

Illustrations

12. Does the illustration help explain the information in the text?
13. Is the illustration a simple line drawing?

* Source: Elizabeth Szudy and Michael Gonzalez Arroyo, *The Right to Understand: Linking Literacy to Health and Safety Training,* ©1994, Labor Occupational Health Program, University of California at Berkeley, 2223 Fulton Way, 4th floor, Berkeley, CA 94720-5120, ph.: (510)642-5507.

14. Is the illustration clear and realistic, with no cartoon figures or complex graphs? In illustrating the human body, have you included as much of the body as possible.

Now review your score for each question. The number "2" represents the best score in all categories. Go back to questions scored with 1 or 0. Make changes to help raise your score and improve your materials.

C. Interest

Interest, like beauty, is in the eye of the beholder. Let us consider three ways to create interest in written materials: writing the materials so they are relevant to trainees' lives and jobs; designing pages that are interesting to look at; and making the materials interesting to use, in other words, as starting points for interesting class activities.

1. Relevance to trainees' lives and jobs

Surveys show workers want training to be immediately relevant to their needs, whether these needs involve workplace activities or personal lives. Experience adds emphasis to this point when trainers learn that most worker-trainees do not want to learn just for the sake of learning, but prefer to cut to the chase and find out what to do to work more productively, more comfortably, or more safely.

A good example of this is the authors' early approach to teaching about the hazards of chemicals. Several OSHA standards require teaching workers about the bad outcomes of exposure to hazardous chemicals. OSHA lists as one of the competencies to be addressed "a knowledge of basic chemistry." We took this to mean, and assumed workers would be interested in, knowing how chemical structure determines chemical behavior. Even though our materials introduced "Mickey Mouse Chemistry" and included unauthorized drawings of the friendly little rodent to ease the fear, chemistry is still chemistry. Class participants were put off — way off — by drawings and explanations of atoms, protons, electrons, etc.

Eventually we figured out that if we started by describing a bad situation (best shown visually by slide or videotape), a situation into which learners could mentally place their fragile bodies, we could then talk about how the chemicals involved had behaved to cause that outcome. Looking at chemical behavior led to discussions of certain chemical properties; how to look them up; how they contributed to the outcome; and how to prevent their behaving in that way. Protons were never mentioned! Starting with job-related reality led to student interest in changing that reality by learning a little — as little as possible in some cases — chemistry.

2. Materials with visual interest

Open up your training manual to any page, and see if you are attracted to look further. Does the page invite you in? Does it promise a good time? Is it visually interesting, informative, and well illustrated? If your answer is not an unqualified "yes" to all of these questions, go back and look at the guidelines and suggestions for writing materials for poor readers. Although class participants with high levels of education (who are experienced at reading dense text to gain information) will wade through training manuals that do not have visual appeal, even they will admit that easy-to-read materials are more appealing. Use the guidelines expressed earlier in this chapter, and those discussed elsewhere in this book for the use of photographs, clip art, and drawings to add visual interest to your pages. If you are writing for inhouse consumption, you will not need the permission to publish cartoons and other drawings that limits the use of such eye-catchers in this book.

3. Materials that are interesting to use

Choose participatory methods from Chapter 5, and style your written materials to lead naturally into being used with these methods. For example, design worksheets for group problem-solving activities that have plenty of room for writing, give clear instructions, and include illustrations. A health effects worksheet to be used with references such as MSDS and the *NIOSH Pocket Guide to Chemical Hazards*, is much more interesting if trainees stick bright orange dots on target organs on a drawing of human organs than if they just list the target organs.

Pages describing appropriate use of fire extinguishers are made brighter with color reproductions of Class A, B, C, and D fire symbols. Material describing Department of Transportation (DOT) or National Fire Protection Association (NFPA) hazard classification systems is enhanced by inserting into the manual a full-color DOT Chart 11 showing all the DOT labels and placards, and providing NFPA labels to stick onto the page at a specified location where classification criteria are listed. When designing problem scenarios for using the *DOT Emergency Response Guidebook*, scan in or insert clip art versions of color placards. Include colorful vendor brochures about respirators, chemical protective clothing, or chemical spill sorbents, and use them to teach compatibility and selection. Scan in photographs of problems, in color if you can, and assign groups to work on solutions.

Use your imagination. Don't try to be formal and impress people with your scholarly publication. This is not the place for that — training is the place to be welcoming, informative, inviting, and interesting.

IV. Summary

Adult literacy has come to be seen as one of the fundamental tools necessary for successful economic performance in industrialized societies. As society becomes more complex and low-skill jobs continue to disappear, concern about adults' ability to use written information continues to rise. Yet we continue to see trainees in all kinds of classes who are unable to use printed materials as a source of information.

Some trainers believe workers who cannot read should not be trained and certified in jobs where reading MSDS, labels, and other hazard information is necessary to protect them from danger. The authors disagree: We think *every* worker has the right to be employed and protected. As trainers, we have to develop alternative methods for hazard awareness; perhaps this must include, at times, a buddy system.

In creating a climate conducive to learning, trainers must establish and repeat the message that they and the people they train have come together to develop safer workplaces with input from everyone concerned. Dispel, by all the means at your disposal, trainees' anxieties concerning their inability to read. This is best done by using your own creativity, and the suggestions in this chapter.

- Write training materials that are visually appealing, user-friendly, and are designed to get your message across rather than merely look good for publication.
- Find and incorporate other means of offering information, in as many different ways as you can.
- Use participatory training methods.

We leave you with good advice from former Vice President Dan Quayle: "Verbosity leads to unclear, inarticulate things."

References

1. 1992 National Adult Literacy Survey. National Center of Education Statistics, U.S. Department of Education.
2. *Education Indicators: An International Perspective*. National Center of Education Statistics, U.S. Department of Education.
3. 1992 National Adult Literacy Survey. Op cit.
4. Ibid.
5. *Literacy, Economy and Society: Results of the First International Adult Literacy Survey*. Organization for Economic Co-operation and Development and Statistics Canada. 1995.
6. *Education Indicators: An International Perspective*. National Center of Education Statistics, U.S. Department of Education.
7. *America's Choice: High Skills or Low Wages! The Report of the Commission on the Skills of the American Workforce*. National Center on Education and the Economy.
8. Hazardous Substances and Public Health 3(3). Agency for Toxic Substances and Disease Registry. August 1993.
9. Szudy, E. and Arroyo, M.G., Evaluating Materials for Readability. *The Right to Understand: Linking Literacy to Health and Safety Training*. Labor Occupational Health Program, University of California at Berkeley, 2223 Fulton Way, 4th floor, Berkeley, CA 94720-5120. Phone 510-642-5507. 1994.

Interview

My name is Rod. I'm the equipment manager for a training program with four instructors, where we do a lot of hands-on, equipment-intensive training. I manage the equipment for various types of training, making sure the equipment is in working order by helping to design, purchase, build, and maintain it. Each course has its own style, and each calls for something different. I work off a schedule or worklist that tells what's needed for the course. When we train out of town, I do the setups for outside exercises and man one of the training stations. After the scenario is over and the class goes inside, I do the cleanup and put the equipment back in working order.

I help a lot of different people, and none of the instructors have the same personality or go about things in the same way. One is a sort of last-minute person; he works off auto-pilot and just knows what he's going to do. Sometimes he thinks I'm thinking like he does and he doesn't tell me what he needs, but if he finds I haven't done something he just works around it. He's easy to please. Another always plans well ahead, and is always prepared. Makes a list, checks it twice; writes a schedule of what she wants done, where she wants it, when she wants it. She's real particular about detail.

Everything I get from a third instructor is written down in detail and I can just follow his list. It's a no-brainer — he just needs a body. It helps me because I can plan my time for the things he wants me to do. If no classes are going on, I have his list of things I can be doing. My job consists of so many different things, it makes it easier if I have a list. The differences in people, and the way they care about what we do, are what I like about my job.

I'm Kecia, and I'm part of the same support staff. I set up for the classroom, and support the instructors with all the materials they need. I want to make sure everything is in place for the instructors so they don't have to worry about whether everything is there. I enjoy seeing them satisfied. You have to enjoy what you do to do a good job, and I really enjoy having everything in order for the instructors so they can just put their mind on the individuals they have to teach.

What I need from trainers is time. Give me enough information and time to get copies made, pack, ship boxes — give me all the information about what I need to do far enough in advance so I can do it. It's a team effort.

chapter five

The adult learner:
characteristics and methods

Adults are different from children in almost every way, including the ways in which they learn. Standard methods used by school teachers are not the best ways to train adults in the workplace, but many trainers use them because these are the only methods they have experienced. If we understand adult learners — their characteristics, wants, and needs — we can do a much better job of training workers.

I. Characteristics of adult learners

Trainers and teachers of adults agree on characteristics of this unique population. Many adult learners share the following characteristics.

- They expect the instructor to be well prepared for class.
- They want to be treated as peers, not talked down to.
- They have life experience and job experience they want others to value.
- They may lack self-confidence after having been out of school for quite a while.
- They expect the material to have direct and immediate ties to factors important to them.
- They may be tired from work, and appreciate teaching methods that add interest and a sense of liveliness, change of pace, or humor
- They probably are not used to sitting and listening for long periods without doing something active.

When planning and delivering training for adults, keep these characteristics in mind. Respect for trainees and their emotional and physical well-being will show in the training you do.

II. Adult Training Methods

A number of different training methods is effective with adult learners, and should be mixed to provide interest and variety in your classes. The attention span of the average adult is said to be about 20 minutes, so a good rule of thumb is to expect adult learners to sit still and listen no longer than 20 minutes at a time.

There are two basic considerations in choosing training methods. The first is your purpose. What do you want people to get out of the training session? The stated desired outcomes, written as your goals and objectives, are the driving force for what you will teach. The second consideration in choosing training methods is you; your personality, talents, and abilities.

Each training method will be discussed in detail in this chapter. All have advantages and disadvantages to be considered in selecting them, and guidelines for using them effectively.

A. Presentation techniques

If you want people to...

- get new facts or information they don't have,
- get an overview of the problem or issue, or
- hear a logical point of view presented clearly,

use the presentation techniques listed here.

- Lecture
- Panel
- Videotape
- Slides

To work well, presentation techniques should be sandwiched between methods that involve participants. The trainer must use discussion or "do-it" techniques that get participants involved. For example, a video presentation should be preceded by questions to stimulate interest in watching the tape, and followed by a discussion of the major facts or ideas in the tape. Find a way to make presentations dynamic (Figure 5.1). Presentation techniques are described on the next several pages, and summarized in the table at the end of the chapter.

1. Lecture presentations

A lecture is a formal presentation, usually by an expert, in which the trainer talks and the workers listen. It may, and should, be brightened by audiovisual aids, which will be discussed at the end of this section.

Figure 5.1 Make your presentations dynamic to capture attention.

   a. Advantages. A lecture provides facts and information to many people at once, and they all hear the same thing. When a lot of technical information must be taught, this is the fastest and most certain way to teach it. The only other way is to have trainee groups dig information out of written materials and try to share it or teach it to each other; this is difficult and very time consuming, and can lead to errors.

   A lecture is the easiest training method to plan and control. If the trainer has good notes, and knows the subject well, all he has to do is talk. There are no interruptions or digressions unless he permits them. Lecture is well suited for presenting new information to a large group when there is little time. The trainer can practice the lecture as many times as needed to avoid simply reading pages to the class, which they will hate.

   b. Disadvantages. Lecture can be, and often is, boring (Figure 5.2). It permits only one-way communication to passive trainees. Retention of information often is low, and a trainer cannot judge the degree of understanding just by looking at the class members. A trainer once enjoyed direct attention from a trainee as he rested his head on his upright hand and covered one eye. She learned later that the open eye was glass and the covered eye was closed. The trainee was asleep.

   c. Using relevant examples. A good trainer is always scheming about how to gain and hold trainees' interest. Lecturing puts you at a disadvantage in this area, but you can overcome it. The only topics for which you should even consider lecturing are those that are scientific for technical enough that an expert is required to explain them. Be very stingy with your definition of what those topics are — don't underestimate trainees' abilities to glean information on their own if you provide the right resources.

Figure 5.2 Many trainees find lecture presentations boring unless the trainer makes them exciting.

Still, you will find that occasionally you need to lecture. Let's look at two different ways to present information about the toxicological effects of chemicals. One uses a well-known example — drinking and hangover. The goal of this session is to promote the use of protective measures to prevent chemical exposure by increasing trainees' basic understanding of how chemicals interact with their bodies. The learning objectives are as follows.

The learner will be able to:

- Explain how chemicals interact with human organ systems once they contact or enter the body;
- Look up chemicals and their target organs and state how certain chemicals have specific actions on particular organs;
- Identify symptoms of overexposure in order to know when overexposure occurs; and
- Provide good reasons why prevention of exposure is more protective than attempting to stop toxic effects once exposure has occurred.

In Table 5.1, you see two ways of lecturing to achieve these learning objectives. The supporting audiovisual aids for each option are as follows. For option one, the instructor uses a stack of prepared, black-and-white overhead transparencies, defining the terms in the outline. For option two, aids include a flip chart and markers (for writing down class responses); a flannel board with a colorful life-size set of human internal anatomy add-ons; stick-on words showing the conversion of ethanol and methanol; yellow arrows for target organs; and red arrows for routes of entry.

Table 5.1 Two Outlines for Teaching Toxicology: Lecture and Participatory

Lecture Outline: Option 1	Lecture Outline: Option 2

Toxicity
  Definition
  Dose-response relationship
Exposure Limits
  Laboratory studies LD_{50}, acute,
    chronic, extrapolation to humans
  Epidemiology studies criteria,
    statistical analysis
  Terminology
    PEL, TLV, STEL C, IDLH, LD_{50},
    LC_{50}, TD_{LO}, LC_{LO}, ALARA,
    mutagen, teratogen, neurotoxin,
    hepatotoxin, nephrotoxin,
    carcinogen
Routes of Entry
  Contact and absorption
  Diseases, chemicals, skin
    anatomy, and physiology
  Inhalation
  Diseases, chemicals, respiratory
    system anatomy, and physiology
  Ingestion
  Diseases, chemicals, digestive
    system anatomy, and physiology
  Metabolism
  Liver enzymes, synergism,
    potentiation, antagonism,
    lipophilic metabolites, excretion
Assessing Exposure Symptoms
Medical surveillance and monitoring
	Ask class: What happens to your buddy
	  when he drinks too much?
	  (Write answers on flip chart)
	These are symptoms of overexposure to
	  ethanol.
	What organs are involved? These are
	  "target organs" of ethanol.
	  (Place yellow arrows)
	What organ do people talk about that
	  eventually fails in alcoholics? (Yellow
	  arrow)
	How does the alcohol enter your body?
	  (Place red arrow) What about when you
	  work with alcohols? (Place red arrows on
	  skin and lungs)
	Explain how alcohol gets from skin, gut,
	  lungs to brain, liver.
	Discuss difference between two alcohols:
	  ethanol and methanol.
	ethanol→acetaldehyde→acetic acid
	methan.→formaldehyde→formic acid
	  (pickles optic nerve) (ant toxin)
	"→" are liver enzyme actions: you
	  cannot stop or change them.
	How anti-drinking pills work
	Why can certain people drink you under
	  the table with no apparent effects?
	Differences between people:
	  genetics, size, fitness, enzyme levels,
	    interactions with workplace exposure,
	    hobbies, lifestyle, medicines
	"Degreaser's Flush" to illustrate
	  synergism; "out of enzymes"
	How to suspect you've been exposed, and
	  what to do

    d. War stories.   In lecture or panel, or when using any method where the trainer has the floor, you must avoid "war stories." Personal examples are good when they are brief and illustrate the topic. Case studies are a useful form of "here's one that really happened." But trainers who use too many personal examples, or drag them on too long or, worst of all, lie about them, should be flogged with a remote control. Trainees get tired of hearing personal war stores, and they will talk unkindly about you on their breaks.

2. Audiovisual aids

Presentations can be much improved by good audiovisual aids. Trainees still are passive recipients of information, but their interest is stimulated by pictures, color, and movement.

a. Overhead transparencies. Overheads are projected on a screen located at the front of the room, either behind or to the side of the trainer. Good overheads have the same characteristics as good posters (Figure 5.3).

- The text is large enough to be legible from the back of the room.
- The amount of text is limited; the overhead is not too full of information.
- Color is used to enhance visual interest.

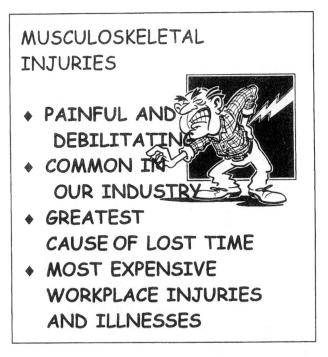

Figure 5.3 A good overhead transparency has graphics, color, limited text, and large lettering.

The number of overheads should be limited. Trainees are put off by a large stack of overheads and begin to focus on the number remaining instead of the trainer's presentation. Also, the trainer with too many overheads is tied to the projector and cannot move around. Overheads used wisely will accomplish one or several of the following purposes:

- Serve as an outline for the presentation so the trainer can get back to his preplanned presentation and the trainees can see where he is and where he is going.
- Show relationships, proportions, and relative importance through the use of graphs and charts.
- Include drawings or photographs to serve as illustrations of the trainer's points.
- Demonstrate how to do, or how *not* to do, a task safely.

b. Slides.   Turn your photographs into slides and you have a set of audiovisual aids that can be arranged to suit your needs. Slides purchased from photo vendors are very expensive, but those you shoot are not. There are also CD-ROM slide sets that can be purchased and turned into slides if you have the right equipment, or your local photo developer may have some available. Digital cameras feed the picture right into your computer for making scenario sheets or overhead transparencies.

c. Videotapes.   Videos have their good use in training, but according to workers some trainers rely on them too heavily. Hundreds of vendors vie for your training dollars in exchange for their products, and some of their products are quite good. Others are not suitable for your topic or your audience; some even contain errors that can cost your trainees their safety. Before you buy a training video, take advantage of the preview process and look for suitability to your workplace and your trainees. Be sure they show the safety rules and attitudes you want to demonstrate.

If you use a video, take the time to write a pre-video questionnaire that stimulates your trainees to watch it. Then find a monitor that is large enough to be easily viewed from the back of the room, or invest in a video projector. These projectors are great! They are expensive, but they turn videotapes into large and compelling training aids. They also have a computer connections for showing presentations.

d. Demonstration board.   Standing wooden boards can be used for topics where you want to show a portion of an actual set-up for an electrical circuit, plumbing connections, or other three-dimensional items that can be attached to the board. For example, they are good for showing applications of lockout/tagout procedures.

e. Flannel board.   An old-fashioned flannel board like the one your vacation Bible school teacher used can be an effective training aid. At a local teachers' supply store you can buy a board, or if you travel, buy a yard of felt at a fabric store and drape it over the flip chart. Purchase felt stick-ons (if you can find them) or make them yourself, and back with felt tape.

One trainer found an accurate, colorful set of human organ systems at a school supply store, and made construction paper arrows to use for routes of entry and to target organ identification. A good artist with a set of magic markers can draw almost anything on white felt, which is self-attaching to the flannel board. The flannel board is useful for sorting words or tasks into their proper order, like the steps in a decision or action process.

f. Computer-based slides.   These slides, when projected onto a screen, are informative and colorful. Unlike other kinds of graphics, they can include motion and sound. See Chapter 9.

3. Panel

Several speakers on a panel present points of view about a topic. Some panelists discuss the topic with each other and answer questions from the audience. The panel should be limited to three members and speaking time to 10 minutes each. Time should be available for discussion and questions.

a. Advantages.   Facts are presented from several points of view. There is some cross-discussion, and some trainee participation if questions are permitted.

b. Disadvantages.   Panels can be boring if they are just a series of lectures, and are primarily one-way communication. There is no chance for everyone to analyze and discuss the topic. If the chairman does not take good control, panelists talk too long or wander off the subject. Sometimes they even argue among themselves, which may or may not provide information for the trainees.

B. Input or discussion techniques

If you want people to...

- understand a new idea or approach in depth,
- relate it to their own experience or situation, or
- accept a controversial or unwelcome idea,

use the input techniques listed here.

- Group discussion
- Brainstorming
- Buzz groups
- Case studies
- Role playing
- Questionnaire

To work well, input techniques require careful organization and guidance by the trainer. Summary and focus must be provided. Ask, "What did we learn from this?" For example, discussion on any subject needs to be guided through the main aspects of that subject by the trainer's questions.

It is up to the trainer to keep the discussion on track and to move it on when one aspect has been covered. This is not always easy, especially for the new trainer.

A number of input and discussion techniques are discussed on the next several pages, and summarized in the table at the end of the chapter.

1. Group discussion

All the members of the class actively take part in a discussion. A discussion should be informal and flexible, developing in accordance with the group's level of understanding, but must be kept on-topic and moving in the desired direction by the trainer.

a. Advantages. Discussions are good for training in an area where understanding, skills, attitudes, and actions are important. They stimulate active participation and offer everyone a voice. Sometimes even argumentative discussions provide insight into a problem.

b. Disadvantages. A discussion can be difficult to guide and control. The trainer's skill cannot always balance the input, since some trainees talk too much and some do not seem to want to take part.

c. Guidelines. Your goal is to get input from a large number of class members. One way to even out the input in a group discussion is to move around. When someone in the left front corner of the room wants to talk too much, walk among the tables toward the rear of the room. Get past him so that you are not facing him, and focus on others.

A lack of reinforcement sometimes serves to stop people talking too much. When a rambler takes a breath, make a noncommittal comment like "okay" or "hmm," then move toward others and ask what they think. Don't put timid non-discussers on the spot by asking them to talk, but if you know one of them has had a relevant experience you might ask him to share it.

d. Play the devil. It's okay to play "devil's advocate" if trainees are not responding. Make an opinionated and inflammatory statement to get people stirred up. You do not actually have to believe the statement to use it as a stimulus to discussion. Example: "Nobody here cares about safety. All workers care about is making more money."

2. Brainstorming

The entire group responds to a question posed by the trainer. Each suggestion is written down on a flip chart, with as little alteration as possible. When a page is full, it is taped onto the wall. Ideas can then be sorted or grouped and discussed by the class. Another way to get written ideas is to give each person or small group several pieces of paper and ask them to write one idea in large marker print on each. Post these, and they can easily be rearranged into groups. Brainstorming must be followed by a summary, which may include a grouping of the various answers to show directions taken by the trainees.

Brainstorming requires a skilled, practiced leader who can think on her feet, listen well, and write fast. It is very helpful to have an assistant to write and tape the sheets, or rearrange the papers. This method can be used in small or very large groups with a good leader.

a. Advantages. Everyone has a chance to be actively involved in seeking answers. All answers are considered, and all are valued equality. Only during the summary are values placed on any of the ideas, and it is the group rather than the trainer who does this.

b. Disadvantages. Brainstorming can be hard to control. If trainees get really involved, answers come faster than they can be recorded and the trainer must keep her wits about her. If the trainer is not lively, ideas may be slow in coming.

c. Guidelines. The trainer never stops to teach, or correct, or discuss a suggestion. No judgments are made about any suggestion until the discussion and summary. Every idea is a good one, and is written down. If an idea is to be removed or rewritten, the group will do it after all ideas are in. If you are in a large group, use a remote microphone and repeat what each person says so that everyone can hear each suggestion.

3. Buzz groups

A large group breaks into several smaller groups, and each group is given a problem to solve or question to answer (Figure 5.4). Give each group a sheet of flip chart paper on which you have written the question at the top, and a marker to write in a specified number of answers large enough for the class to read when they report back to the entire group. You can give them smaller pieces of paper for individual ideas if you plan to group them in the summary.

This activity works best when the question does not have merely factual answers. For example, in a class of paper mill workers, a lively room full of buzz groups discussed questions such as "Why are some workers apathetic about safety?" and "How can you convince other people to work safely?" The same groups, discussing questions about material safety data sheet information, were much less animated and involved.

a. Advantages. Everyone gets involved in the discussion. Buzz groups help trainees to analyze and synthesize information. If the questions posed are relevant to the topic and the group, many of the trainees will have experience that is valuable in solving the problem.

Figure 5.4 Good buzz group questions will set the room buzzing with ideas.

b.   Disadvantages.   It is easy for small groups to get off the topic and shoot the breeze among themselves. Unless they are given a time limit and reminded of it as discussion progresses, some of them will not attack the problem in a timely manner.

c.   Guidelines.   The trainer must give a good introduction and a careful summary. Questions must be carefully thought out. Move from group to see if they understood the assignment and are progressing, but don't interfere if you are not needed. A time limit for report-back should be set, and groups given ample warning as time moves toward the limit.

4. Case studies

The trainer presents a realistic case, and the group suggests how it might be handled. Often it is a case that has already been decided or completed and the "answer" can be provided after discussion. The case should be realistic, and one that people can identify with. It should be complex enough to have several possible solutions. Give enough facts to provide adequate food for thought.

For example, one class considered a safety grievance filed by workers with documented allergic dermatitis who felt the company had not responded to their verbal complaints. A class in labor law considered a case where a worker had been fired because she participated in a same-sex marriage ceremony after being instructed by her supervisor not to do so.

a.   Advantages.   Discussion of a case involves every class member. It can provide several points of view and several reasonable solutions. Many cases have an emotional component, including some injury or worker rights issue, and trainees get very involved in the discussion.

b.   Disadvantages.   Case discussion takes time. The group may argue (which is not always a bad thing), or may get off track into a discussion of similar cases that do not enhance the discussion. It is not always possible to locate or invent a case that illustrates a certain topic.

c. *Guidelines.* If the case is real, be sure you know how it actually was resolved. If you make it up, have someone you have confidence in go over it with you for all possible resolutions so you will be prepared for anything trainees come up with.

5. Role play

Two or more people act out a situation that relates to a topic the group is learning or discussing. It must be carefully introduced, and is always followed by a discussion by the entire class. Role play requires special talents on the part of the trainer, who must write a situation that is complex enough to be engaging, but through which trainees can find their way. This is a good method for helping people understand other points of view, and is more suited to emotional or attitude topics than purely technical ones.

A group of workers and their managers may be asked to do a job safety analysis, and then play a worker and a supervisor negotiating the changes the worker requested. In some of the groups, a manager plays the role of a worker and the worker is a manager. The reversed roles allow people in the class to view a topic from a different perspective. Workers appear to be quite familiar with management's responses to their requests.

a. *Advantages.* Role play stimulates an analysis of different points of view. It involves everyone emotionally, and brings out feelings and attitudes. It can be fun to watch, as well as to do. Playing roles helps people practice actions they learn in class and want to take back into the workplace.

b. *Disadvantages.* This method takes longer than others if it is well done. The trainer's preparation is extensive, and the trainer must have a sense of group members' personalities to assign parts. This is one of the most difficult methods to do well, and should not be attempted "cold" without prior practice in role playing.

c. *Guidelines.* These are difficult to do well. There is a good chance your players will be wooden and just stand there, or turn the role play into a silly farce. (Silliness is not always a bad thing, if it's brief and breaks building tension). The best chance of success comes when the role differences are based on attitudes and opinions, like bargaining issues, rather than facts, such as OSHA regulations. If you don't think your players are creative enough to play their roles without one, write a simple script.

6. Questionnaire

A questionnaire is a brief, simple set of questions that are to be answered before information is presented. Its purpose is to stimulate interest in the information so trainees will pay attention to it. The same questions may be answered again after the presentation. Sometimes questionnaires are used at the beginning of a class to assess trainees' knowledge and to set the parameters of the class.

   a.   Advantages.   This is a good way to get people to think about information to be presented, especially if the questions require specific answers that are found in the following presentation. The questions can be as general or as specific as the leader wants.

   b.   Disadvantages.   People with poor literacy skills cannot respond to a written questionnaire, and a visual or verbal one must be designed. The questionnaire will take longer than you think it will, so design it to be short.
   The questionnaire in Figure 5.5 is shown on an overhead transparency and used to stimulate interest just prior to the showing of a videotape that contains the answers to the questions.

C.   "Do it" techniques

   If you want people to...

   • build skills,
   • put their learning to work, or
   • experience "doing it,"

use the hands-on techniques listed here.

   • Problem-solving workshops
   • Small group activity method
   • Risk mapping
   • Action plans
   • Projects
   • Hands-on practice
   • Exercises
   • Simulations

   To work well, "do it" techniques need introductory guidelines and plenty of feedback from the trainer. For example, in learning how to perform a chemical hazard assessment, each student should use references as directed by the trainer. Trainers should provide feedback on important topics, such as the good points of certain references, and how to avoid common mistakes.
   A number of "do it" training methods are described on the following pages, and summarized in the table at the end of the chapter.

QUESTIONNAIRE

1. What color is the placard on this page?

2. What can happen to a DANGEROUS WHEN WET labeled chemical if it gets wet?

3. Red labels and placards show what primary hazard?

4. When a truck carries a DANGEROUS placards, what does this tell you is in the truck?

5. How does the four-digit number on some tanker truck placards help you in an incident?

6. In the NFPA hazard class system, what does the number in the blue diamond tell you?

7. What kinds of chemicals require CORROSIVE labels and placards?

8. What kinds of chemicals are shipped and stored in plastic drums?

9. What is the most serious warning word on a pesticide container?

10. What kinds of places would you look for PCB?

Figure 5.5 The questionnaire stimulates interest in the videotape to follow.

1. Problem-solving workshop

Small groups of trainees respond to a problem or scenario designed to require them to use reference materials and judgment in making decisions. The scenario must be carefully designed, and the necessary references available. If the scenario or worksheet covers a technical topic, the necessary prerequisite training must precede the workshop. Often a problem-solving workshop is used to help trainees apply and summarize technical training.

a. Advantages.   The workshop is a good way to teach many topics, especially the use of manuals and references. It is a good way to lead trainees to make decisions in realistic scenarios when that is the desired outcome of training. It actively involves trainees.

b. Disadvantages. Workshops are time consuming. Some students will not participate if the group is dominated by strong or very knowledge-able members. If this seems to be a problem in a class, the trainer must be careful how trainees are grouped. Sometimes placing all your engineers in one group is better than spacing them among the groups, and leaves the other groups free to come to their own solutions.

c. Guidelines. Provide all the information trainees will need to solve the problem. These may be in the form of fact sheets in their training man-uals, or references available in the back of the room, or individual pocket reference guides you provide for each individual.

Some groups need to learn to solve problems and some do not. For example, trainees who work for a state environmental agency and respond to transportation-related chemical releases should spend considerable train-ing time-solving problems, as this is what they do on the job. A group of process operators who will evacuate during a chemical spill would not benefit from the same kind of practice.

2. Small group activity method

The small group activity method is used by some trainers to structure the entire course as a problem-solving workshop. Worker trainees sit at round tables in groups of eight or fewer, with a facilitator (who is also a worker) at each table. The activities are conducted as information searches, guided by carefully written questions, problems, or case studies. Groups of workers seek answers and solutions from written material, provided as information sheets.

This method is used by at least one international industrial union as the course model and sole method, as they believe it empowers workers to seek their own answers and make their own decisions. No outside "experts" such as toxicologists, safety professionals, or industrial hygienists do any teaching in the courses.

a. Advantages. The workers do, indeed, learn how to glean informa-tion from written materials of various kinds. They are highly involved in the course, and no potentially boring training methods, such as lecture, are used. Trainees should leave the course with a feeling of competence and empowerment.

b. Disadvantages. This method takes considerably more time than some others, as answers must be located and there are no members in the groups who are highly educated in the topics. For the same reason, some trainers worry that wrong or incomplete information may result from the group work, with no one available to correct the errors.

c. Guidelines. Provide all the information required for the groups to complete the activity. This may be in the form of fact sheets, large classroom references, or individual pocket references. Provide a mechanism on the worksheet for dealing with questions the group cannot answer by asking, "Where will you go for further information?" and providing a list of sources.

3. Risk mapping

Risk mapping is a conceptual framework for the investigation of workplace hazards. The instructor opens the session by asking the class to create a map of the workplace on which are indicated all the health and safety hazards they know exist there. This leads to the discussion of a number of related topics, including everything you want to teach about workplace safety. For each hazard, you can use one of the methods in this chapter to teach recognition, prevention, control, correction, regulatory compliance, and a number of related topics.

The risk map is not simply a visual aid; it is the structure on which the entire training is built. For that reason, it must be accurate and include considerable detail.

a. Advantages. This is an interesting way to train, and it involves all the trainees from the very beginning of the class. It is specific to their workplace.

b. Disadvantages. Risk mapping is time consuming, but most good methods are. It requires good advance preparation on the part of the instructor to ensure that the map made by the class is accurate and complete.

c. Guidelines. Set this up with a good explanation of what you want to do, but don't start it with a lecture. Don't teach the topics until they evolve from the risk map. Have plenty of information ready to deal with all kinds of hazards. Have available a large, laminated site map or a large map and plastic overlays, with washable color markers. Flexibility in the order of and relationship between topics will be necessary.

4. Action plans

Action plans can be designed around any topic. They help workers lay the groundwork for setting out on a course of action that will result in a safer workplace.

- The workers focus on specific problems that need attention.
- They prioritize their list of problems, and select the top two for action.
- They write objectives for their plans and a time line for achieving them.
- They decide on direct, step-by-step routes for taking action on the problem.
- They design follow-up activities.

RISK CHART: CHEMICAL EMERGENCY PREPAREDNESS
Use a red dot sticker to indicate problem areas. Use a green dot sticker to indicate areas that are okay.

_____The company emergency plan is known to all workers and everyone knows his or her role in an emergency.

_____All workers recognize the chemical emergency signal.

_____Hazard Communication or First Responder Awareness Level training has taught all workers to recognize and identify a chemical spill and the potential hazards of all chemicals in the area.

_____All hazardous chemical containers, including pipes, are labeled.

_____People who respond to and clean up chemical spills are trained for the tasks they are expected to do.

_____Evacuation and/or response drills are carried out at least once a year.

ACTION PLAN WORKSHEET
CHEMICAL EMERGENCY PREPAREDNESS

From the Risk Chart, select two (2) red dots that represent priority health and safety problems that concern you most in your work area.

RED DOT
1:_____

RED DOT
2:_____

For each of the red dot problems, what do YOU need to do to solve the problem? What does the COMPANY need to do to solve the problem?

Red dot 1: What I need to do: Red dot 2: What I need to do:
What the company needs to do: What the company needs to do:

Figure 5.6 Action planning sets the stage for safety changes.

An example of an action plan is shown in Figure 5.6. On the actual worksheet, space is provided to answer the questions. On the second worksheet, the trainee is asked to think about obstacles to action, and plan how they may be overcome. He sets a timeline, and writes down specific steps for naming individuals to talk to, what to say, how to measure success, and how to publicize the accomplishment of the goals.

   a. Advantages. One safety manager likes action plans because they "teach workers how to get me the information I need. It is not effective to come to me and say, 'Everybody's getting sick back there. You need to get us all respirators.'" One union-based trainer likes action plans because they turn workers into safety advocates with a definite plan. Many workers like them because of the guidance they provide in helping to generate changes.

   b. Disadvantages. Trainees go back to work ready to make changes. Their supervisors or managers may consider this a disadvantage, but as the trainer you probably will not. Teach the advantages of persistence without hostile confrontation.

   c. Guidelines. Group trainees together by work groups or plant departments. Trainees with common problems can work together for common results. Workers who have been generally griping will need guidance to focus on specific problems and tasks. They will also need help in carrying out the planned actions in the face of resistance when they go back to work. When action plans are used in training, the trainer may want to keep a copy and follow up at three- or six-month intervals to see how the trainees are progressing with their plans. The actual plan, or at least the format, will be useful to workers who are frustrated by their inability to improve a particular aspect of safety and health in the workplace.

   5. Projects

Trainees work together to accomplish a task set by the trainer. The task may be mental or physical, but should be complex enough that it requires planning, doing, and presenting to the class. A train-the-trainer class, for example, can plan a team teaching effort, present it to the class, and get feedback from members of the class. A chemical safety class might be asked to catalog the chemicals in their work areas and evaluate the hazards and the container labeling system, or select sorbents for different areas of the plant (Figure 5.7).

   a. Advantages. Trainees work together, and can utilize the skills of the group members in a variety of ways. In the train-the-trainer example given above, not all members of the team must teach but they all must be involved in some way in the project. The biggest advantage of projects is the involvement of the participants, leading to more learning and better retention.

   b. Disadvantages. Not everyone will share the responsibility equally. There is no way for the trainer to ensure equal work, except by encouragement. As with all hands-on or small group activities, projects are more time-consuming than some of the presentation methods.

Figure 5.7 Trainees work on a group project.

   c. Guidelines.   Don't overload people with too much structure. Let them set goals and plan the project. If there is to be a worksheet, a planning sheet or an evaluation sheet, make it very specific for the information you want them to use. For example, practicing trainers should list their topic, their method(s), their reasons for choosing both, their training materials, and the main point they want to get across. If there is time and you have taught them to do it, have them write goals and learning objectives for the session.

6. Hands-on practice

Individuals and groups use tools, equipment, and other devices that will be used on the job (Figure 5.8). The practice tasks are simulated, progressing to actual tasks as trainees work.

   a. Advantages.   This is the best way for trainees to gain physical competence. Hands-on activities bring several human senses into the learning process, which enhances retention on the part of the trainee. As a rule, trainees love hands-on practice, so you are providing what they want. When students evaluate your course, they will thank you.

   b. Disadvantages.   Hands-on practice is very time consuming. It requires equipment, so it can be expensive. It requires space, either indoors

Figure 5.8 Trainees extricate a confined space "victim."

or out. The instructor-to-student ratio must be lower than in most other methods, so more instructors are needed.

   c. *Guidelines.*   The task design should be as realistic as possible. The trainer designing the tasks should always consider the safety of the participants. Safety will probably demand a larger number of instructors. The National Institute of Environmental Health Sciences, (NIEHS) which funds a number of worker training programs across the nation, recommends the following instructor: trainee ratio for hands-on activities.

   1:10   Levels C and D personal protective equipment
   1:5   Levels A and B personal protective equipment

   If outdoor activities are conducted during weather extremes, especially heat, use all best practices for preventing heat stress and prepare for emergencies. Require physical fitness evaluations or self-evaluations when trainees will be asked to perform physically demanding tasks, and reserve the right to ask a trainee to sit out an activity. Use a checklist for documenting competency for the tasks trainees are doing.

   7. Exercises

Exercises are hands-on practice with a script. They provide a setting in which trainees will use several skills developed through hands-on practice or group workshops. For example, trainees who practiced wearing a self-contained breathing apparatus yesterday while walking around the field, and this morning applied plugs and patches to "leaking" pipes that were empty, will

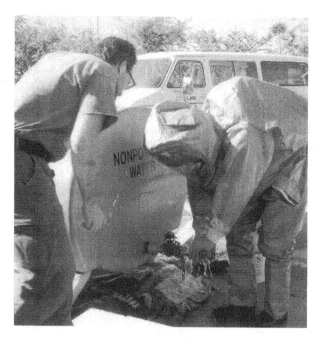

Figure 5.9 Several hands-on activities come together in the exercise, closely monitored by the trainer.

put on SCBAs and chemical protective suits this afternoon and repair the same pipes as water flows through them (Figure 5.9)

   a. Advantages.   Exercises are part of a natural progression toward doing the job well and safely. Prerequisite skills are practiced singly until they are mastered so that when trainees put them together they are not difficult. Also, trainees really like exercises.

   b. Disadvantages.   Time is the big disadvantage. There is also a need for equipment, space, and enough trainers for safety and helpful feedback.

   c. Guidelines.   Take the time to get it right. The primary criterion for exercises is that you need the right equipment at the right place at the right time, and all of it must work. Just in case, have someone standing by at each station to "gofer" missing or replacement items. Make a setup list with a diagram for each exercise to help you or your field assistant. Allow plenty of time for setup, and have enough rain suits on hand that you can continue outdoor activities even if it rains. Schedule time for a full critique and feedback session, including videotapes of the exercise if you feel they will be helpful.

Figure 5.10 Simulations can be complex, requiring a great deal of planning by trainers.

The safety guidelines listed for hands-on activities apply here as well. It is a good idea to have CPR-trained people on hand if the exercise is strenuous.

8. Simulations

Exercises become simulations when the script includes all the problems that may occur together in a given job. For example, the culminating exercise in a week-long Hazmat Technician course could be a simulated truck wreck with leaking "chemicals" (simulated by water) in a setting where the class forms its own incident management structure and solves the problem (Figure 5.10)

    a. Advantages. With simulations, you have all the advantages of exercises plus the decision-making requirements that are placed on the group. The class comes together to make all the decisions and perform all the necessary tasks, and this leaves them with self-confidence and competency in taking care of problems.

    b. Disadvantages. The same as for exercises, only more so. You may even have to contact the local fire and law enforcement agencies so they do not respond to the neighbors' calls about your smoke bomb or the sight of trainees in moon suits.

    c. Guidelines. Simulations require a great deal of preparation, and often demand the cooperation of a number of department or agencies. Engage the help of experienced people to serve as evaluators and controllers, and give them special shirts or vests so that they can enter restricted areas. If the controllers are to provide information to the participants as the simulation progresses, give them complete scripts of the information they are to deliver, indicating when and to whom it is to be provided.

If you have never designed a simulation, get help from those who have, or start small. Experience is crucial in knowing what to expect and what will work. If you can videotape an exercise or simulation, trainees will find their own mistakes when you play it back to them. Some knowledge of the learning objectives is required for the camera operator, since you do not want 4 hours of people walking from task to task but prefer to have 30 minutes of the important actions. If trainees are using radios in a simulation, place a radio on the camera so transmissions are included in the video tape.

Members of the class may want copies of the videotape. People love to watch themselves on television. If no company policies are violated by sharing the tape, do it. Showing the tape to friends and family will reinforce the learning process. If you can't spend the time or money to make copies, a trainee will probably volunteer to do it.

D. Group-building techniques

If you want people to…
- get to know each other,
- feel part of a group and build solidarity, or
- share ideas and experiences freely,

use the group building techniques listed here.

- Icebreaker
- Small Groups
- Team Projects
- Problem-Solving Workshops
- Open Group Discussions

For group-building exercises to work well, the trainer should tell the group in the beginning that they will learn more and work together better if they know one another. For some activities, you may want people to talk to the people they work with every day about a common question or problem, or mixing people from different shifts or plant areas might work better for the exchange of ideas. Decide this in advance.

Some of these group-building methods have been discussed in detail in other sections of this chapter, and will not be repeated here. All are summarized in the table at the end of the chapter.

1. Icebreaker

Class members exchange information about themselves with each other. A report of the information follows, introducing all members to everyone. The leader should adapt the information sought to the topic of the training.

Icebreakers should force people to get up and move around the room, not just talk to the person next to them.

The icebreaker in Figure 5.11 has been used in different settings. The questions could be changed to fit any training topic. After the trainer introduces himself and the course, participants are given a sheet containing the same descriptions you see on the screen. They are to find a different person in the class to fit each description, and the trainer(s) can be included in the eligible group. When their search is completed, the trainer asks for a name to match each description on the screen. As each name is written in, that person is asked to stand up and tell the class how the description fits him or her, as well as mentioning the workplace and special health and safety interests. Any trainees who do not get listed on the board are asked to introduce themselves. Try to cut off long speeches, especially the presentation of workplace problems; those can be discussed later. This icebreaker can be lengthened with additional questions, or adapted to other topics by changing the questions.

a. *Advantages.* The icebreaker gets people moving around and interacting with each other. It is done early in the class to facilitate working together. It sets a tone for an enjoyable, flexible class and shows the trainer has a sense of humor. Trainers need all the good karma they can get at the beginning of a class, and an enthusiastic icebreaker is one way to get it.

b. *Disadvantages.* Some people are shy, and are uncomfortable with icebreakers. The trainer must carefully word the questions so that they do not ask for personal or embarrassing information, unless you know the group well and are sure they will not be offended.

c. *More icebreakers.* The design of icebreakers is limited only by your imagination. Several that we have seen or used with good results are described below.

Fun Facts. This can be used if you have access to trainees in advance of the class; or if not, you can use it on the second day. Ask each participant to write down a "fun fact" about him- or herself. Give examples, such as "Shot an eight-point buck last year," or "Can eat two pounds of chocolate in one day," or "Once flew in a hot-air balloon." Be sure their names are on their papers. The trainer creates a matching exercise with names down one side and facts (not in the same order) down the other, copies enough for the class, and lets participants draw lines to try to match them (Figure 5.12). Confessions follow.

What's My Line? As class begins, each person writes large on a piece of paper a job or hobby or activity in which he or she participates. Mix them up, then hold them up and let the class decide whom they describe; that person tapes the paper onto his or her front. When everyone has a sign, they admit which one is really theirs and tape the accurate ones to their shirts until the next break. These give participants something to talk to each about

FIND SOMEONE WHO . . .

Has been involved in a chemical-related accident.

Is on a joint safety committee.

Can tell you his/her respirator size.

Is hazmat trained above the awareness level.

Has been involved in safety for 20 years or more.

Has been involved in safety for 1 year or less.

Is a trained accident investigator.

Always works safety.

Has no hazardous chemicals in his/her workplace.

Has read his/her company's written emergency plan.

THE RULES:
Different person for each question
First and last name (facility and/or company optional)

Figure 5.11 Change the descriptions in the "Find Someone Who..." icebreaker to fit the group and the topics.

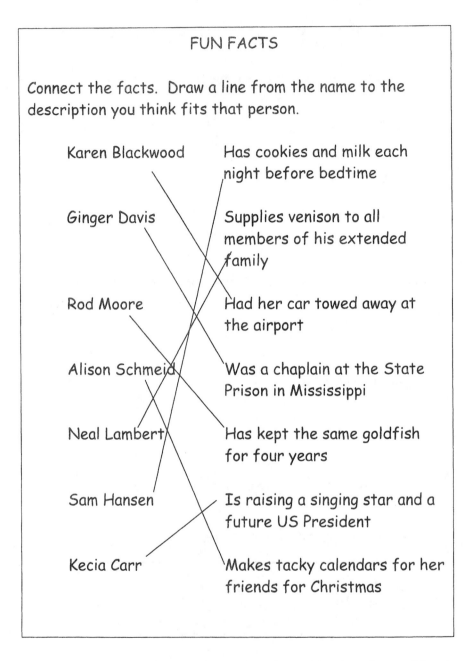

Figure 5.12 The "Fun Facts" icebreaker requires advance input from class participants, or can be used on the second day.

on the break as they get acquainted. This is time consuming, so it is best done in a small class.

Paper Bag. This icebreaker can be used when you want the group to bond in some way, usually in order to work together on a long-term project. It will be effective only if the group will be meeting long enough to begin to know each other, perhaps for several days. Send instructions to participants in advance so that they can bring the bag with them, as the items require thought and need to be brought from home. Instructions should read: "Bring a paper bag to class on the morning of the first day. On the outside of the bag, attach items that describe or illustrate who you are at work. On the inside, place items that show who you are when you are away from work."

Individuals are asked to show and describe the items, starting with their work persona. Items illustrating each participant's other self can be shown immediately following the work items, or later to introduce a different topic, depending on the structure of the class or meeting. Or, if the objective is to group people by jobs or by personal interests, participants can be asked to move into groups based on items inside or outside of the bag. The facilitator can determine what sorts of items match the topic or the objectives of the session and ask for grouping by a particular item or class of item (such as fish hooks and hunting arrow points, or a softball and cycling gloves).

Last Menu. In small groups, participants who assume they only have one day to live write up a menu for their last meal. Each member of the group must agree to eat each item. The meal must include an appetizer, an entrée, a drink, a vegetable or salad, and a dessert. Groups will find this difficult. Then announce that each group must send a representative up front or out of the room with their menus, and the representatives must compile their menus into a meal for the entire class. They write the final menu on a flip chart the class can't see, which the trainer then shows to get reactions from class members. This icebreaker can be used when the topic is negotiating, compromise, or representing your co-workers or union members. It sometimes teaches tolerance for medical or religious differences, depending on the unique natures of the participants.

2. Other group-building techniques

The other group-building techniques listed in this section — team projects, small group exercises, problem-solving workshops, and open group discussions — have already been discussed. These participatory training methods serve your agenda in several ways.

III. Guidelines for dividing people into groups

It is important to have a plan for grouping trainees. Many a trainer has realized too late that to say "Okay, everybody get into groups" is a signal for chaos if you haven't already made it very clear who is to go where. Describe your plan before you ever say the words "Divide into groups."

Some activities work better when trainees are grouped with the people from their work areas, and some are best when trainees are mixed for a variety of ideas. In a class that includes more than one group activity, rearrange the groups for different exercises so that an expanded exchange of ideas is possible. The more people an individual groups with, the more likely he is to learn a variety of possible tips or solutions to workplace problems. Here are some ideas for signaling how groups will be divided.

A. Colored dots

Before class or during a break, stick colored dots purchased from an office supply store onto training manuals, chairs, or table places. Prepare a flip chart page to serve as a map — draw large colored circles that match the dot colors. Trainees go to the part of the room indicated on the map, and there they meet their group. Number of colors = number of groups.

B. Count off

Start at a corner and have students count off one through four (for four groups), or five, or whatever number of groups you want. Write on the flip chart where the ones will meet, and the twos, etc. Some people who aren't paying attention will get their count wrong, and you always have to say before you start that they need to remember their own number.

C. Name tag symbols

If you are using name tags, stick or draw a symbol on each name tag. At the center of each table, place a symbol matching a group of name tags. This works well for large groups when you don't want to take time to count off.

D. Turn the tables

If trainees are arranged on only one side of long, narrow tables, half of them can turn around and meet with the people on the other side of the table behind them. This is a method we like to use for the first grouping when the class has not yet loosened up and is still worried about what we are going to do to them. People feel safer when they don't have to leave their chosen seats.

IV. Trainer characteristics and training methods

In the descriptions of training methods, we have included some guidelines for trainers. It is important when choosing which methods to use to consider your own personal "advantages and disadvantages" and match them to the method. It is admirable to work in new, untried methods as you become more proficient in training; however, it is recommended that you start with methods you are pretty sure you can handle. Asking the following questions of yourself will help to direct you in your choices.

A. Do you know the material well?

For example, if you are teaching people to use the DOT Emergency Response Guidebook, are you good at using all its parts? Have you practiced it enough? What is that funny "P" symbol in the latest edition? Try it out on inexperienced friends or family members, and see if you can answer their questions.

B. Are you comfortable in front of people?

If you are not comfortable, use a method that gives you more control, such as lecture, panel, a problem-solving workshop you can plan thoroughly in advance, or a hands-on exercise for which you have enough equipment.

C. Can you control the group?

Can you take control of the group if they wander off the subject, or one person talks too much, or they sit like stumps and don't respond to your questions? Discussion, case, and some icebreakers require this. Ability to do it has two parts; your basic personality, and practice in doing it.

D. Can you hold ideas and summarize?

Can you keep your head and summarize what several people are saying at once, and write it down as you go? This is important for buzz groups, and for summarizing small group activities.

E. What kind of assistance will you have?

If you are trying to keep more than four groups on task as they do buzz groups or problem-solving exercises, things will go better if you have someone to help with some of the groups. This is especially true if the trainees are not highly educated, not experienced in group work and idea synthesis, or include people who talk a lot and tend to get off track.

F. Is there time to work through the exercise?

If you are using scenario-based training, do you have the time and patience to completely work through the entire exercise in advance? If you don't, surprise glitches are guaranteed.

G. *How many people will help?*

Will you have plenty of help with hands-on exercises? These take constant and complete attention from the trainer. More than one group requires more than one trainer, to ensure good instruction and safety.

V. Summary

This has been a long chapter because it is an important subject — probably the most important one of all. Choosing a variety of training methods and learning to use them well will make your training exciting and effective. Everyone will benefit if you take this chapter to heart.

- Your trainees will feel respected, be interested in what you are teaching, and accomplish the learning objectives.
- You will enjoy training.
- The evaluations of your classes will improve.
- Training will be effective and everyone, including the trainer, will have fun.

All the adult training methods discussed in this chapter are summarized in Table 5.2. Why not add just one new method to your next training session? Pick a class when you have enough preparation time to follow all the guidelines. Practice the delivery or the exercise thoroughly before you try it. If you are unsure whether it will work, you might even tell the class in advance, "This is a pilot course for a new idea. I hope you will let me know how effective you thought it was, and how I can make it better next time." Then, on the evaluation form, ask this specific question again and leave space for answers.

Table 5.2 Summary of Adult Training Methods

Method	Description	Advantages	Disadvantages	Comments
Lecture	Formal presentation usually by an expert	Gives facts and information to many people at once; easier to arrange and control	One-way communication; students passive; retention of information is low; can't judge degree of understanding	Particularly suited for presenting new information to a large group when there is little time
Panel	Several speakers present points of view about topic, discuss with each other and audience	Presents facts from several points of view; some cross-discussion, some participation	Primarily one-way; no chance for everyone to analyze and discuss for themselves	Good chairman is important; limit to 3 members and speaking time to 10 minutes each; leave time for discussion

Table 5.2 Summary of Adult Training Methods (continued)

Method	Description	Advantages	Disadvantages	Comments
Discussion	Group members actively take part in examining problem by talking about it in an organized manner	Good for training in areas where understanding, skills, attitudes, action are important; stimulates active participation; builds solidarity	Takes longer; difficult to guide and control; some students don't take part, while some talk too much	Needs well prepared, skilled leader; points should be summarized frequently; informal and flexible, developing with group's level of understanding
Problem-solving Workshop	Groups respond to problem or scenario designed to require them to use reference materials and judgment in making decisions	Good way to teach use of manuals and references; actively involves trainees; permits synthesis of information	Time consuming; some students will not participate if their group is dominated by strong or very knowledgeable members	Scenario must be carefully designed and necessary references available; if materials are technical, train on their use prior to workshop
Icebreaker	Class members exchange specified information with classmates, report on information gained	Gets people moving around the room, interacting with each other; provides interesting format for introductions	Some people are shy and are uncomfortable with icebreakers; some are lazy and don't want to get up	Leader should be innovative in setting the task; works better if people have to get up and move around, interact with others
Questionnaire	Brief written questions are answered before information is presented, and again after	Good for getting people to think about information about to be presented	People with poor literacy skills cannot use written questionnaires	Questions can be as general or as specific as leader wants
Case	Leader presents a case with the facts for group to suggest how it can be handled	Involves everyone; provides several points of view; several possible solutions	Takes time; group may fall into argument or get off track into discussions of similar cases	Case should be real or at least realistic, and one people can identify with; give enough facts to make it realistic, complex

Table 5.2 Summary of Adult Training Methods (continued)

Method	Description	Advantages	Disadvantages	Comments
Buzz Groups	Large group breaks into smaller groups; each group discusses problem or question and reports back to entire group	Involves everyone in discussion; helps class analyze and synthesize	Needs room suitable for several small groups; may get off track; may misunderstand the assignment unless monitored	Leader must monitor groups to see that they understand the topic and focus on it; requires good introduction and careful summary
Hands-on Practice	Individuals use tools, equipment, other devices that will be used on the job; tasks are actual or simulated	Best way for trainees to gain physical competence; proven to increase retention of material; trainees like it	Requires equipment and time; also requires space, indoors or out; instructor: student ratio must be low	Task design should be as realistic as possible; safety considerations are a factor
Role playing	Two or more people act out a situation that relates to a problem the group is discussing; response to role play follows	Stimulates analysis of different viewpoints; involves everyone; brings out feelings and attitudes; helps people understand other points of view and get better group understanding	Takes more time; group may worry more about "acting" than about the problem involved; some people just can't do this	Needs to be carefully introduced, very carefully designed, and followed by discussion

Table 5.2 Summary of Adult Training Methods (continued)

Method	Description	Advantages	Disadvantages	Comments
Brainstorming	Entire group responds to question posed by leader; responses are written on flip chart; pages taped onto walls and possibly rearranged; summary follows	Gets everyone actively involved in seeking answers; permits all ideas to be considered	Can be hard to control	Requires a skilled practiced leader who can think on feet, listen well, and write fast (or has a helper to write)

Good luck. We know all these methods work if they are applied to the right material, and if the instructor is comfortable and prepared. We know because we have used them all in actual training sessions with real workers.

Interview

I'm Kenny, and I'm a full time trainer, in this same job for eight years. I was doing environmental assessment work but wanted to move back into the area where I'd just completed a master's degree, industrial hygiene. I planned to stay a couple of years, but as I started doing it more and more I realized training was very interesting to me.

I prepare for training by trying to understand not just the narrow topic but also the context in which it sits. I'm not comfortable at all with rehearsing a presentation. I prefer to have a rather sparse outline from which I operate and carry on more of a conversation. That enables me to talk to people, read their responses both visually and verbally, then feed on that and carry the topic through in the direction that seems most appropriate. I end up with the key points in outline form and ideas of the details that would be appropriate.

I like to provide as much written materials to trainees as I can. A lot of information transfer occurs in class, but it's also helpful to have a good reference to come away with so when you get back into the workplace, days or months down the road, you have a good place to go back and refresh yourself or dig a little deeper into the topic.

I've learned from the people I do training with; more of a relaxed style from one; from another, to keep things well organized and write down details that might escape me in front of a group. From others I've gotten a better sense of how to "work a crowd," how to relate to a group of people so that they're not allowed to distance themselves mentally from me and from the class, to keep them engaged.

I come into class with the attitude that I'm not necessarily smarter than the folks I'm talking to; I've simply read more. I believe if I can understand it, then surely I can help you understand it, because if I can do it, so can you. I have an interest in making sure people in the class understand, so I'm careful to look for clues that I'm not keeping them engaged, or I'm confusing them or going over their heads. I have the tendency to ask questions that, even if they solicit obvious answers, get people talking and get communication going.

At the risk of sounding clichéd, people don't care how much you know unless they know how much you care. That really applies in training. It will be immediately apparent to whomever you are training if it's just a job for you. To the extent that you can relate to the class and have a discussion or a dialog with the trainees, you'll have that much more success. The key to effectively communicating information is having an interest in the people you're communicating to.

chapter six

Training over the generation gap

Generation X has arrived in the workplace! A whole new group of young adults, with different values, life experiences, and communication styles, have been thrown into the vast melting pot of the training population. This makes the blend a little different from what trainers have ever faced before. To deliver quality training to the new workforce, trainers must understand why the new generation differs so drastically from previous generations. In this chapter, we will try to answer two important questions.

- What characteristics do the members of Generation X possess that set them apart, and how do they relate to learning style?
- When all the current generations are represented in the classroom, how will trainers handle their differences?

Understanding these questions will allow trainers to step outside the realm of personal experience and offer training that is more effective by putting themselves into the world of the trainees.

It has been said that Generation X's life experiences, learning styles, and educational expectations are so different from those of previous generations that traditional training techniques don't stand a chance. But wait a minute! Should trainers design and redesign training materials around this one outlandish group? Before we grumble too much, perhaps we should consider whether training geared toward this generation may actually be better for training workers representing *all* generations.

I. Defining the generations

Three generations currently are represented in the workplace; the Silent Generation, the Baby Boomers, and Generation X. Are they really that different from each other?

The Silent Generation was born or grew up during the Great Depression. When they entered the workforce in the 1950s and 1960s, they did so during a time of economic growth. This made the job market plentiful and the chance for advancement and growth within the company "a done deal." For the Silent Generation, this meant job and economic security. If you started working for a company as a young person, you could realistically see yourself retiring from that same job.

Baby Boomers were the generation that grew up during the economic prosperity of the 1960s. They are the largest demographic group in America and still hold the majority in the workplace. During their teenage and college years, the Boomers experienced the sexual revolution and enlightenment of the 1960s. Like the Silent Generation before them, when the Boomers were ready to enter the workforce, they entered during a time of economic growth. Much of that growth was directly attributable to the sheer number of Boomers entering the workforce. This group lived the excesses of the 1980s. They became better known as the "hippie-*cum*-yuppie" generation. Unfortunately, when reality hit, it hit this generation hard. The late 1980s and early 1990s ushered in the economic downsizing of large corporations in which large numbers of upper and middle managers lost their jobs, accompanied by the deletion of entry level positions. Corporations began to slim down in order to survive and many Boomers lost their jobs in the process.

Generation Xers grew up in the 1970s and 1980s, shadowed by the accomplishments of the Boomers. This new generation represents over 40 million American workers born between 1961 and 1981. When the Xers came of age, they did so during a time of corporate downsizing, increasing inflation and credit debt, and declining wages and benefits. They believe they are the lost or forgotten generation no one cares about, hence the signature "X." They feel like they are a day late and a dollar short, because when it became time for them to enter the workforce there were no good jobs available: The job market was tougher than it had been for either of the previous two generations.

Mounting national debt, the failing Social Security system, the AIDS epidemic, and the ever-tightening job market give Generation Xers a sense of hopelessness. They feel like they are being left to clean up the mess of the overindulgence of the previous generations. They have come to the realization that they will be the first generation not to do as well as their parents.

Xers are better known in some circles as the Twenty-something Generation, the Baby Busters (because they are the smallest demographic group), or the 13th Generation or the Thirteeners (because they are the 13th generation since the ratification of the Constitution). But whatever we call them, in order to train them we need to understand why their learning styles are so different from the older workers we train. Let's look at some of the characteristics that set the members of Generation X apart from previous generations.

II. Who are the Xers?

If we consider how society has changed in the last two decades, it is easy to recognize how change has affected the Xers.

A. Latch-key kids

The children of this generation were the first to grow up with both parents working and/or in school furthering their educations. Many grew up in single-parent homes, products of the ever-increasing divorce rate. In 1950, 7% of U.S. children lived in single parent homes; today, 24% do. This is the first generation of "latch-key" kids who were left unsupervised in the home while their parents were at work or school.

B. Cynics

Xers tend to mistrust large institutions. Growing up, they witnessed the media coverage of multiple government scandals, and watched as their parents were laid off because of corporate downsizing and buyouts. In 1980, a CEO earned 41 times the average factory worker's pay. By 1997, that same CEO position paid over 285 times the average factory worker's pay. This is very disheartening to the Xers, since they are the average factory workers. Xers don't have to be told that there is no such thing as lifelong employment. They never expected it.

C. Want it now!

Society has taught Xers that they can have what they want when they want it. Devices such as automatic teller machines, pagers, cellular phones, microwaves, and remote controls have conditioned them to immediate gratification (Figure 6.1). Delayed gratification is not unknown; Xers simply see no need for it.

D. Technoliterate

Xers grew up with computers in their homes and schools, so they are very technoliterate. Don't forget — this is the generation that grew up with Space Invaders and Nintendo. They are the kids Dad asks to program the VCR. Not only is this generation very familiar with computer technology, but technology has been changing at such a fast pace that they have had to keep up with it. This is nothing new to them, it's just how it always has been.

E. Lifelong learners

Xers consider themselves lifelong learners. They view training as an opportunity to grow, to add to the resume. Xers seek jobs that will offer them continuing education. And because they don't expect to stay at the same

Figure 6.1   Xers will even make calls in the middle of your class!

company until retirement, they value training that will push them ahead in the job market.

F.   *Want a reason*

Xers like to see clearly defined goals or learning objectives. They want to know how new information will relate to them or to their job duties. They want to know *why* they must learn something, before they will learn *how*. When learning, they want the work they do in a training session to be meaningful. Having read Chapter 5, you know this is characteristic of most adult learners; certainly it is true of the majority of workers.

G. *Focused*

Xers are focused. They hate busy work. If the relationship between a task and the learning objective is not clearly defined, Xers get frustrated. Their minds will wander quickly, and they'll be pondering some other intricate problem of life, like what they're having for lunch, instead of concentrating on the task at hand.

H. *Xers are fun!*

If you keep them stimulated, Xers will add a great deal of fun to your class. The Xer who managed to sneak a huge fright wig into his moon suit for a spill response simulation exited the suit wearing the wig and announced "We've identified the chemical! It's Rogaine!" The young guys who yawn loudly when a lecture runs too long burst out of the starting gate when physical action is required, and demonstrate they have been listening to every word. The trainee who dons huge plastic ears behind the back of his favorite big-eared instructor (who loves it, and he knew that) brings laughter to the classroom. The firefighters who don't mind saying they relish the respect and family atmosphere of their co-workers set a cooperative tone for the class.

III. How do Xers learn?

Being "latch-key" kids made the Xers a very independent and self-reliant generation. They also developed good problem solving skills very early on. On the up side, this makes adult Xers very good at getting the job done on their own. On the down side, because they did not always get the amount of feedback they craved from their parents, they expect it from their bosses. When given a job assignment, they want independence in doing the task (they don't want a boss hanging over them telling them how to do each step) but they do like feedback on how they are doing. In the classroom, they expect the same of trainers. They want to be given a project; they want that project to relate the new information to their life experiences; they want to be given the elbow room to work on the project, but they want feedback from the trainers on how they are doing. Sounds complicated, but it's really not.

A. Demand control

It is important to Xers to be able to control their own learning experience. They are used to managing their own time. They are independent. Because of this, if you try to force-feed them training, they will resist. If an atmosphere can be created that gives the Xers a sense of control over their learning experience, it will be more meaningful. Also, giving Xers as many options as possible about locations and times of training to attend will help them gain the control they seek.

B. Seek stimulation and change

Xers expect immediate gratification. This means adult Xers are very responsive in the workplace. They crave stimulation and if they don't get it, they become bored very quickly. What does this mean in the classroom? That adult Xers tune out very quickly if they don't get the stimulation required to keep them on track. One trainer watched an Xer pull out a newspaper during a lecture, and thought "How rude!" Later, she talked to the trainee and realized he was interested in the material but bored with the presentation.

Xers are conditioned to rapid change. Change doesn't bother this group like it has the previous generations. They are not afraid of new technology; in fact, they enjoy the challenge it presents. They adapt and adjust as needed. Computer-based training doesn't scare this group off. In fact, because of the Xer's independence and the computer's ability to give immediate feedback, in some situations a computer-based learning experience may be a good change of pace from other training methods.

C. Parallel thinking

Xers have the ability to focus on more than one idea at a time, which allows them to assimilate lots of information quickly. This ability, called "parallel thinking," is one reason Xers love computers. If you watch young adults using computers, you will see they click around a lot. They can control their path, absorb loads of information, and start or stop at will. They love this.

Most TV commercials have five or six things going on at the same time. These commercials are designed to keep the Xer's attention while delivering multiple messages at once. This drives Boomers up the wall, but Xers are used to it. Popular publications such as *Wired* or even *USA Today* have designed their layouts around this idea of "parallel thinking." These publications combine larger text, photos, graphics, charts, and cartoons on one page. When designing training materials for classes that include Xers, do the same. Make your written materials eye-catching. Use larger text with lots of white space. Don't make them guess what you're trying to say. Spit out your message as simply as possible. Highlight key words or phrases. Xers are scanners. The more scanner-friendly your training materials, the better. Use charts, graphs, photos, etc., when appropriate, to present the information in a format that is different from the written word. Make sure that this format makes it easier to pick up main ideas at a glance.

IV. Training strategies

Trainers, the majority of whom are not members of Generation X, may find themselves a little angry with Xers and their needs. Boomer trainers tend to perceive Xers as arrogant, lazy, and having short attention spans. Most Xers really aren't arrogant. They are just self-reliant and independent. This self-reliance and independence makes this generation appear very confident; those who don't understand the Xer view this confidence as arrogance. Get over the defensiveness this group provokes in some trainers, and design strategies to improve their learning. These same methods also are effective with older workers.

A. Keep it short

The fact is that Xers do have relatively short attention spans. Most sources put that attention span between 15 to 20 minutes. This means that trainers should keep lecture times down to 20 minutes, tops. Follow this same rule when selecting videos. If you don't, you will lose Xers (Figure 6.2). It is best to mix lengths of lecture with some type of participatory training method (Figure 6.3). Be creative. Keep learning as experiential as possible. Once you have their attention, keep it by making the training experience meaningful and fun. Make the learning environment one that will facilitate ongoing learning.

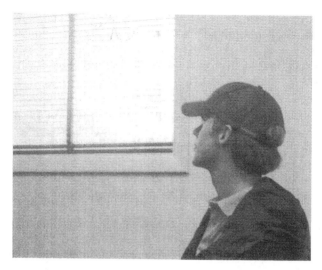

Figure 6.2   Young trainees get bored easily, and their attention wanders . . .

Figure 6.3   . . . but put them in a participatory training setting and they shine.

Trainers sometimes view Xers as being lazy. Xer traits may lead to this perception.

- Xers do have relatively short attention spans. If you lose their attention, it is hard to get it back. Once you've lost them, they don't participate.
- Before Xers will learn how to do something, they want to know why they must know this and how it will benefit them. If you don't show

this first by laying out objectives, you will have trouble getting and keeping their attention.

- Xers hate busy work. If they feel like the trainer is just wasting time in the classroom, they will not participate. If they don't see the relevance of an exercise, they will not complete it. Make sure the training message is clear, to the point, and that it is something they can understand. Make sure that the relevance of each exercise is known up front. Don't make them wait until the end for you to tie information together.

If these suggestions seem familiar, it's because they are described in chapter five where they are recommended for all adult learners.

B. Keep it focused

In Chapter 2, we discussed writing outcome-oriented objectives. Focusing on outcomes helps Xers put information to work. These trainees need to know not only why they must learn something, but also how to put the information to work. Focusing on the *how* will make your class more meaningful to the Xer. For example, teach them why a container needs to be grounded and bonded before transferring flammable liquids, but also show the correct methods for grounding and bonding. In one training organization, the *why* is taught in the classroom by lecture and demonstration, showing very vividly what can happen if containers are not properly grounded and bonded. Following the demonstration, the class participates in a hands-on activity in which they are taught *how* to ground and bond containers. Several scenarios are used, along with different types of containers and different grounding and bonding equipment. The trainees have to make decisions not only about the correct equipment to use, but also have to show knowledge of the methodology of grounding and bonding.

C. Keep them engaged

With Xers in your class, it is important to choose training methods that will allow trainees to learn by experience (Figure 6.4). If they do something, they will be more likely to remember it. Techniques such as buzz groups, case studies, role-play, and small group activity not only allow the trainee to gain knowledge by experience but also to build on the knowledge and experiences of others in their group. These techniques engage trainees and can help them expand their thinking and reasoning processes. These techniques also help the trainer in that other members of the trainees group are offering the feedback that the Xers need to stay involved.

Figure 6.4 Learning by doing is important for students of all ages, and vital for training Generation X.

D. *Make it challenging*

Provide challenges for Xers. Offer realistic scenarios in which the answers to the problems are not so obvious. Design exercises that use multiple skills, even if they are skills that the trainees have learned previously. In doing this, you allow the trainees to incorporate new information and skills with their existing knowledge base and also practice skills they already have learned. A knowledgeable instructor offers this quote, "Repetition is the mother of skill." The more you practice at something, the better you'll get.

As a trainer, don't be afraid to try new things. Diversify your teaching materials and methods as much as possible. All learners, especially Generation Xers, must be engaged for learning to occur.

V. *Summary*

Xers are different; no doubt about it. The gap between the generations is related to radically different life experiences of the individuals in those generations. Trainers who target Generation X will be targeting the characteristics of most of today's learners, no matter what their age or generation. Using teaching methods and strategies that facilitate learning in the Xers will also make learning more meaningful and fun to the Silent Generation and the Boomers.

Interview

I'm Alan, and I'm a trainer. The biggest change I've seen in my 11 years as a trainer has to do with the technology available to trainers. I started out making transparencies with a machine that stamped letters one by one onto a tape strip, which I would then tape to paper, copy, touch up with white-out, and copy onto transparencies. Now we have software and graphics for computer presentations, and I've gotten more techno-literate.

I think I spend more time than a lot of other trainers in preparing to train, because that is what it takes for me to do a good job. I think about how I want to pitch things based on what the predominant audience in a class will be; for example, firefighters or industrial workers. I make sure slides or transparencies are in the right order and I'm familiar with things so I won't get stuck, or end up out on a limb, or go into something at the wrong part of a lesson. In some cases, I read over the text material the teaching aids track, and for field exercises, make sure everything we need is in order. Probably my most significant contribution is that I bring aspects of practicality to our program, creating things to train with "where the rubber meets the road," where you go from the technical and hypothetical to the concrete.

I've learned from things I've seen in other trainers that I thought were good — like people who are comfortable with their situation or audience, and have a good enough grasp of the material and the way they are presenting it so they are not intimidated by questions or comments. Or bad things — like people who do a presentation using transparencies and just put up a transparency and read down the list of items, rather than using it as an outline to elaborate on.

I get ideas spontaneously, particularly if I have an obligation to come up with something new. What's important is to write it down — that's why I always have an index card in my pocket, and a pen. I also get ideas by observing how other people do things: sometimes I can bootstrap off their idea, modify it, and link it up with something else to solve whatever my problem is. We develop new material for annual refresher courses for several of our courses, which forces us constantly to come up new ideas. That's an advantage, and we can use the update courses kind of like a proving ground to try out new ideas and, if they work well, roll them back into the regular courses as ways to achieve the learning objectives for those courses.

Doing this kind of work is really a higher calling, since how you do your job determines whether somebody gets hurt. Others may get fired, or not have a good sales year, but what we do involves an obligation and responsibility.

chapter seven

Participatory training: designing activities and workshops

We can't say enough about participatory training. We can exhort, plead, and present facts to support the superiority of participation over passivity. Most effectively, we can let you try it and come to your own conclusions. This chapter will include examples of hands-on and interactive training that have been used by effective trainers. It is by no means a complete list, but will serve to stimulate your imagination and help you get started on your own designs.

Participatory training includes any activities in which trainees are expected and encouraged to *do something*. They are actively engaged in talking, planning, solving, analyzing, synthesizing, or physical motion; often they are doing several of these at once.

Trainers who choose participatory training methods gain so much, but, as with all choices, give up something in the bargain. The advantages of participatory training considerably outweigh the disadvantages.

I. The disadvantages

Let's go ahead and get the disadvantages out of the way, because they are minor and can be easily overcome.

A. Time

Participatory training takes more time. You should plan to take at least twice, and probably three times, the amount of time you would use to simply tell trainees how to do something. Information you can state in 10 minutes may take student groups an hour to distill from their references and discussions using the small group activity method (explained in Chapter 5). Buzz groups and other small-group activities take time to explain, move people into, do, and follow with reports of group conclusions.

You can show a class a set of plugs for leaking drums in a minute or two. You can project a drawing of the plug in place in a drum during the same few minutes. To set up hoses running into drums, divide the class into teams to plug and patch the holes, have them use plugs and patches, and then clean up after them will take at least two hours. But which way more effectively teaches people to plug leaks?

To shorten the time required for a field exercise, form teams before going outside, give clear instructions illustrated by a flip-chart map, and have the teams move about at will, completing each station as it becomes available rather than moving in order. In classroom problem-solving exercises, use the following guidelines to keep time wasted (defined as time off-task) to a minimum.

- Give instructions before moving people into groups. Don't try to do it while they are moving and getting acquainted.
- Announce the time planned for the exercises at the beginning, and state the time to be finished for the people who will forget what time you started.
- Make an announcement when there are ten, and then five, minutes left.
- Shut up and leave the class alone while they work. If you continue to interrupt their work with clarifications and further directions, they will get irritated. They will also lose track of where they were — and some of them won't hear you.
- Plan for abbreviated report-backs, if report-backs are required for the exercise.

There are ways to reduce the amount of time used in reporting back. Let's assume each group has a different scenario or problem, but all have the same worksheet or report-back form. If the spokesperson for each group reads through her entire form, the reports soon become redundant and time begins to drag. When you pass out the report-back form, have a different set for each group; state that group one should report the conclusions they wrote on the first section or question, group two the second question, and so on. Even though each group had a different scenario, if each spokesperson briefly reads the scenario the class benefits from one answer out of the several on the sheet. Often a spokesperson, who may be a little nervous about talking in front of the group, forgets the instructions and begins to read the whole sheet. When you call on group two, ask for their answer to question two. Politely interrupt if the reporter starts at the beginning, asking, "What conclusions did you reach for question two?" The class will silently thank you for causing the reports to go more quickly, and you will not find yourself drumming your fingers on the table and wishing they would get to the end.

Managing time is the biggest problem when you choose to use participatory methods. The authors have found it is better to leave something out of the class when time is limited than to opt for an all-lecture format in order to get it all in. Once again, remember our credo: You can train workers effectively while everyone, including the trainer, has fun — or at least, stays awake and gets involved.

B. Work

Participatory training requires more work. One group of four instructors arrives at the crack of dawn to set up the "truck wreck" that brings together all the topics of a week-long Hazmat Technician emergency response course. An hour and a half of heavy physical labor gets the four-hour exercise ready. The mess made by the response, dressout, decontamination, and stabilization of the incident leads to another two hours of cleanup by instructors, and a full day of equipment cleanup.

Classroom workshops, too, take more work on the part of the instructor. An example is the ten-station Yellow Drum Exercise (Figure 7.1) used as a

Figure 7.1   The Yellow Drum exercise took a lot of work to construct, but is used to good effect in three different courses.

small-group activity to teach emergency procedures to hazardous waste site workers (also applicable to industrial workers). The OSHA-mandated hazardous waste site worker course requires 40 hours of training in a variety of topics, one of which is responding to emergencies during the assessment and remediation of hazardous waste sites, like Superfund sites and old landfills. Mutiny would result if trainers sat these workers down at a table for five full days — most of them are quite unaccustomed to sitting. A yellow 25-gallon steel drum and an accompanying yellow toolbox contain all the materials for directing small groups to cooperatively find solutions to chemical spill situations. Stations are laid out on tables, with drawings, photographs, tools, plugs and patches, and laminated question sheets. The drum itself has four punctures of different shapes, into which certain plugs will fit. Class participants move from station to station, answering questions on a worksheet included in their training manual.

The trainer who put the Yellow Drum Exercise together spent many hours writing the questions and worksheets, creating art pages, and collecting the necessary tools, equipment, informational brochures, and sorbents for the exercise. Another trainer put just the right holes in the drum. All items are contained inside the drum when it is not being used, so the exercise is convenient to move from room to room, or even to ship.

A very meticulous instructor built several accurate scale models of trucks and rail cars. Hours of assembly time went into the models. When they were completed, the instructor "wrecked" the models, added placards and representations of fire or leaks, and glued the models onto boards with scenery to simulate hazardous materials incidents (Figure 7.2). Scenario descriptions and directions for hazard analysis and/or response actions were included in the exercise, along with student worksheets to be filled out.

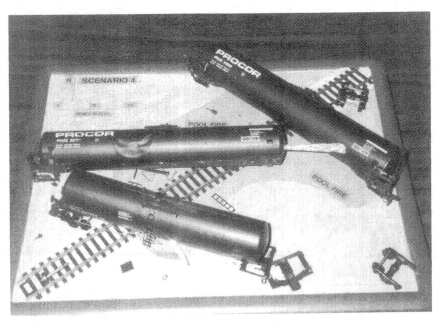

Figure 7.2   Simulated Hazmat incidents include meticulously built models of tank cars or rail cars.

C. Money

Participatory training can cost more money. If you or someone else can use junk or other materials already available, do so. Building devices from odds and ends, junk, and spare parts is quite inexpensive. Contracting for metal-work, carpentry, and welding can run into money. Chapter 8 includes designs for hands-on training devices.

For a number of participatory activities, the only expense involved is development time. With a good computer, a graphics program, and some paper, an imaginative instructor can easily create exercises that engage train-ees' minds, and often their hands.

To get a professional-looking end product, you may have to pay profes-sional prices; perhaps you will decide it is worth it. In a low-budget univer-sity setting, the authors have never been bashful about putting homemade scenarios and training devices in front of trainees. Trainees have appreciated our efforts and some of them, who might have been put off by a more polished approach, seem to prefer our way of doing things. Some trainees have come back later and helped us build things.

II. The advantages

Time and money are in short supply, and most trainers don't need to add to their workload. Why do participatory training? The advantages far out-weigh the disadvantages.

A. Increased learning

First and foremost, trainees learn so much more. According to the New York Literacy Council, adults remember

- 10% of what they read
- 20% of what they hear
- 30% of what they see
- 50% of what they see and hear
- 70% of what they say
- 90% of what they say and do at the same time.

B. Enjoyment

Participatory training is fun! Every trainee will enjoy the class much more. Trainees will write better course evaluations, and will spread the word that the training was great. The trainers will be tired, but they'll have more fun, too. Everybody leaves the class smiling (Figure 7.3).

C. Visibility

Participatory training is visible. If your status as a trainer is elevated by visibility, train outside the classroom where people will see you. Invite the company newsletter or local news photographer to visit at a time when your trainees are doing a hands-on activity, or one that involves spirited discus-

Figure 7.3   Participatory training leads to smiling trainees.

sion and lots of trainees with their mouths open, smiling, and talking. Never underestimate the advantages of showing off your training program. Personal ego inflation is not the point here, so don't be bashful. The more people who hear what good training is taking place, the more good training you will be able to offer the workers you serve.

III. Guidelines for participatory training

We have already covered classroom participatory training methods and provided some advice and comments about their selection and use. The remainder of this chapter will discuss the design and use of workshops and hands-on exercises to accompany and enhance classroom activities.

A. Build skills

Build the workshop step-by-step. Each step is done at a different time, perhaps even on a different day. Be sure each skill required in the workshop has been learned and practiced separately before being combined with other skills in the workshop. The following is an example of a progression of skills for hazardous materials technicians.

Day 2 — Trainees don respirators; do negative and positive fit checks; wear air-purifying full-face masks and self-contained breathing apparatus (SCBA).

Day 3 — Trainees use plug and patch tools and devices on drums leaking water, as well as pipe stands with various kinds of holes but no water (spraying water requires protective clothing) while wearing no protective clothing or respirators. They wear SCBA for 30 minutes while doing light exercise.

Day 4 — Trainees don SCBA and dress out in Level A totally encapsulating chemical protective suits. When they are comfortable, they enter a simulated "hot zone" and perform several patch and plug tasks on leaking drums and pipes.

Day 5 — The class sets up an incident management system (IMS) and responds to a simulated truck wreck or other hazardous materials incident. Several class members enter the hot zone, asses the incident, and bring the spills under control, while the others serve in other roles of the IMS for which the class has prepared them. A post-incident critique follows.

The same kind of progressive skills acquisition is important for classroom participatory activities. For example, teach all the basic information-gathering skills and then ask groups of trainees to fill out hazard assessment worksheets describing two chemicals they work with, or all the chemicals involved in a scenario you provide. Show slides and examples of a structured process for making decisions in emergencies, then ask them to use the process in a group pencil-and-paper exercise which takes them through the steps in a described emergency. Provide examples and explanations of shipping papers and container labels; then ask groups to actually fill out papers, pack a sample, and label a box (Figure 7.4).

B. Practice in advance

Always run through a new workshop or exercise completely before you use it in a class. Invariably, a glitch will be discovered. Practice will allow you to correct any deficiencies in planning or equipment. Try the exercise on friends or family members who lack expertise in the topic — if it makes sense to these beginners, it can be done by newly hired workers.

C. Allow adequate time

The first several times the class does the workshop, give yourself more time than you think you need. If you overestimate, spend the extra time letting trainees discuss the workshop outcomes. If the exercise includes planning or decision-making on the part of the class allow even more time, as their knowledge about the scenario is less than yours. Also, you want to teach them to take their time in hazardous situations. Experienced workers may be in the habit of taking short cuts without proper hazard analysis. Firefighters are famous for rushing in to "put the wet stuff on the red stuff," although recent training innovations have slowed them down in Hazmat response. Other trainees may have learned quick-fix habits you want them to break. Remember that situation analysis and decision making are time consuming.

Figure 7.4   Packing and labeling a box for shipment requires a progression of skills.

Keep in mind that trainees may learn more from other class participants than they do from you. Give them plenty of time to seek input from all members of the group. Discussing common problems may bring out practical solutions from others who have already solved them.

D. Plan for safety

Some of the hands-on exercises include activities that can be dangerous. Consider the activities and the conditions under which you will be doing them, and be very sure you are prepared to handle all potential outcomes. You probably need first aid and CPR training, and perhaps an EMT standing by. Plan to screen out, using whatever means you feel is appropriate, people whose participation in certain exercises you believe to be hazardous. We are deliberately being vague here, as these decisions can be made only by you, your organization, and the employers of the class members.

OSHA relies on the employer to ensure fitness for work. The training tasks should be based on what the worker will be expected to do; therefore, fitness for duty equals fitness for training.

E. Plan for grouping

When dividing the class into groups to do problem-solving exercises, keep certain things in mind. The first thing to remember is to have a plan for doing this (see Chapter 5 for grouping methods).

- Separate the people who know the most about the topic, putting one into each group.
- If you have a slow reader, be sure he is with someone who will take the time to include him.
- If you find certain people are always dominant in a group, ignore the first suggestion above; put the all bossy ones into one group, allowing the others to develop some leadership.
- For some exercises, it is best to put people from the same department or shift together. Do this when you use real workplace examples or scenarios common to a particular department.
- When you divide again for a different exercise, mix people in new ways.

While exercises are being done, wander from group to group and listen. Just asking, "How's it going?" will show your availability if they need your help. If there is an obvious problem, lead them by your questions into overcoming the hurdle themselves.

Whenever the group presents their results from a classroom exercise:

- Suggest they choose a different spokesperson each time.
- After the presentation, ask members of the group other than the spokesperson if they have anything to add; then encourage the rest of the class to ask questions.
- Summarize the presentation very briefly.
- If you disagree with their conclusion, ask them why they decided to do it that way, and question with "What if . . .?"

IV. Examples of participatory training

We will present examples of several training methods that involve the trainees in participatory activities. Examples of the following kinds of participatory training activities are included in this chapter.

- Problem-solving workshop
- Scenario-based workshop
- Classroom stations exercise
- Outdoor stations exercise

A. *Problem-solving workshop*

Design a problem you will describe on paper or set up with equipment or models. Trainees use what they have learned, perhaps including MSDS or other references, to decide how the problem can be effectively solved. For example, you might have groups of three or four participants fill out a worksheet showing the hazards of each chemical in their work area. Information could include chemical properties and how they help predictions of outcomes, health hazards and symptoms of exposure, and considerations of who may be at risk in a spill or fire situation. Provide whatever reference materials will be available in the workplace.

A good problem-solving workshop that applies to a variety of safety and health classes is the job safety analysis (JSA). We introduce the JSA with a videotape, free from the U.S. Mine Safety and Health Administration, that gives instructions and provides a case study. Groups of trainees then choose a job task and, using prepared worksheets, break the task into steps, evaluate the hazards of each step, and suggest preventive measures for each hazard.

In industrial classes, trainers have found that class participants enjoy making group safety posters. They can be assigned to combat a particular safety problem. A lot of thought goes into the posters, and the only tools required are flip chart paper, color markers, and tape to put them on the walls for the ensuing poster contest. Winners get candy (and so does everybody else).

B. *Scenario-based workshop*

Trainees are divided into groups, and each group is given a scenario that could happen in their work area or jurisdiction. The scenario is described in writing, sometimes accompanied by a photograph, gas-tight sample bag, pocket reference, model, or diagram. Trainees use the materials provided to answer questions on the accompanying worksheet (Figure 7.5).

For hazard communication training, DOT Hazmat employees, or OSHA first responder awareness level, scenarios might describe situations that are illuminated by the use of the *DOT Emergency Response Guidebook* or other placard and label guidance references. Scan in labels to print with the scenarios, or use clip art. If a color printer and copier are not available, develop scenarios based on those labels that are black and white (Corrosive, Poison, Class 9), or ask trainees to fill in the label color as part of the worksheet. Give them red, green, or blue markers and let them color the labels. This sounds elementary, but they will enjoy it and it qualifies as a learning experience.

Scenarios can be written to accomplish almost any learning objective, whether it is to analyze hazards, select protective clothing, change behavior, or improve the effectiveness of the safety committee. Worksheet spaces or questions gather facts and information about the situation, and organize thoughts and actions leading to a proposed response.

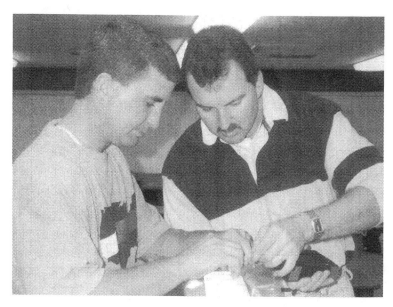

Figure 7.5 Trainees use a photoionization detector to determine the concentration of contaminants in an air bag.

C. Classroom stations workshop

Stations are set up around the room. Trainees are asked to perform a task or answer a question at each station. They can move in any order they want, taking advantage of open stations so they don't have to stand in line (Figure 7.6). Tell trainees they may work together in partners or groups, or can work separately if they prefer. When everyone has completed the worksheet, use the items at each station, plus transparencies and slides where appropriate, to go over each station and discuss the tasks or answers.

Classroom station exercises are a good way to review personal protective clothing and respirators (Figure 7.7). At the stations, trainees might select a cartridge, inspect an air-purifying respirator or a chemical-protective suit for damage, describe the use of a cooling vest, or choose the appropriate protective fabric for use with a specified chemical. Students can learn or review selection, inspection, maintenance, and care by this means.

Store all the items for the PPE exercise in a box. At each station, set out the following.

- One or more items of personal protective equipment
- A plastic-protected sheet stating the problem or asking the question
- Any reference or other item needed to answer
- References to the training manual pages where more information is found

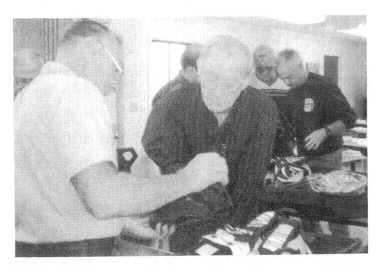

Figure 7.6   The classroom stations workshop stimulates bodies and minds to work.

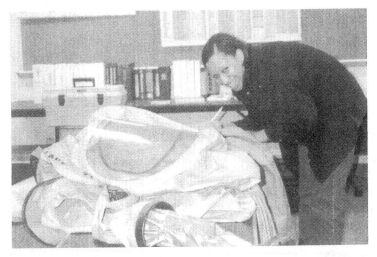

Figure 7.7   Exercises are good ways to teach, or review, personal protective equipment.

Worksheets are handed out before the exercise, or included in the manual if it is in loose-leaf format.

One trainer finishes the personal protective equipment stations exercise by setting up a "PPE store" containing a variety of items, then asks each team to assemble a different level of protection: Level A, Level B, Level C, or Level D. Teams are invited to dress out a team member in the ensemble. Does this take more time than simply telling the class what items are included in each ensemble? Yes. Is it a more interesting and more memorable way to teach levels of protection? Definitely.

D. Outdoor stations workshop

This workshop requires physical work, and generally takes longer at each station than the classroom stations workshop. For this reason, the number of stations may be fewer than in the classroom. Trainees move from station to station in pairs or teams and do the assigned tasks (Figure 7.8). In most outdoor workshops, a trainer needs to be available at each station.

Figure 7.8 Trainees do assigned tasks at several stations. This group is building a runoff dike.

In an emergency response technician's class, one field exercise could include the following stations.

1. Two clean open-top drums with holes. The holes in one drum are fairly regular, but of different sizes, and accommodate various wing-nut plugs. The holes in the other are ragged and irregular, requiring some sort of epoxy or paste patch. Both kinds of plugs and patches

are available, as well as rubber gloves and a trash can. Hoses run into the drums, and can be turned on when needed to leak from the holes.

2. A closed drum approximately 1/3 full of water; an overpack drum; a drum lifter; rollers for inserting one drum into the other; protective gloves.

3. Two pipe stands and assorted pipe clamps. Pipes of various sizes have been cut or drilled to cause a variety of leaks. Pipe joints are also leaking. If the exercise is done without protective clothing, the pipes are dry.

4. A 150-pound cylinder that has been pressurized to simulate leaking, (see Chapter 8 for directions for making this one), and whatever tools and equipment are required to stop the leak

Draw a map of the exercise, plus a list of all the tools and other items that are to be staged at each station. No matter who sets up the exercise, nothing will be forgotten.

E. Combining skills in an exercise

After trainees have learned and practiced several skills that will be used together on a task, set up an exercise in which they combine all the skills into one performance. One of the most enjoyed combined skills exercises used by the authors is the Leak Monster exercise, where pairs of trainees repair a series of leaks in a device built from an old water heater (see Chapter 8).

The combined skills exercise can be as extensive as the trainer wishes, from the simple combinations described earlier in this chapter to a full-blown disaster simulation (Figure 7.9). Whatever the skills required, the exercise will not achieve the learning objectives unless all participants have achieved the prerequisite abilities prior to the exercise.

F. Games

Some trainers like to use games as training activities. (A computer-based Jeopardy-style game is described in Chapter 9.) Most trainees enjoy the games, but some cautions should be mentioned.

- A few people think games are silly and don't like them. We have encountered one such person among hundreds of trainees.
- The trainer has to be careful to keep it light and avoid heavy competition.
- Some people are so competitive that they get angry if their team loses, no matter how much fun everyone else is having. We have seen one bitter loser in 12 years of training.
- Games must be played between groups, not individuals, and the group should consult before the designated spokesperson says the answer. That way, no one is put on the spot for an answer.

Figure 7.9 Simulations require great amounts of planning and work, but they are worth it.

- Games are very time consuming to write; software is available for some of them (Chapter 9).
- The literacy levels and lifestyles of the participants should match the requirements of the game; for example, people who watch television are familiar with Jeopardy, while people who don't read for pleasure probably have never played Trivial Pursuits. People who enjoy crossword puzzles tend to be fairly highly educated. Inexpensive software is available to make crossword puzzles.

Games are good to use in refresher courses, since they let trainers and participants find out how much they remember. They are a good way to go over topics already covered in a new and interesting way. Normally they are not designed as the original presentation method for information.

V. Summary

As you can see, the sky's the limit on designing participatory activities. These exercises, like everything else in training, should be planned, designed, and conducted to aid in the achievement of the written learning objectives. Once that criterion is met, and the issues of time, money, and help for the trainer are handled, participatory training is the best, and some trainers say the only, way to train effectively.

Participatory training is so effective in generating results and so much fun for the participants, both trainers and trainees, that you will not want to go back to the old way once you try it. Participatory training is more than a new fad or buzzword — it's a method that truly lives up to its claims.

Interview

I'm Lisa, and I'm a trainer. I got into training around eight years ago by accident. My background is in chemistry, and I was the only technical person at my company, so the safety and health fell into my lap. In the last three years I've been a full-time trainer, I've gotten more relaxed, more open, and more able to start a conversation with the class. That draws them out and allows them to interact.

To prepare to do training I read over the lesson plan, if there is one, and go through the teaching materials and make my own mental notes. I do not like to walk into a classroom without having looked at the material before. The first few times I taught new subjects, I would start two weeks early and write an outline to study to get the flow of the material. If I didn't, I would get to a spot where I'd become stumped.

I think people appreciate the chemistry experiments I do. It shows abstract concepts, and they can see what I'm talking about, like flash point, or lower explosive limit, or chemical reactivity. In the workforce, they don't mix a lot of reactive things together on purpose in an uncontrolled atmosphere, and the only time a bad thing happens is when something accidentally goes together. If I can show them with small quantities what can happen, they have a better understanding of chemical safety. I get demonstrations from other instructors, or chemistry books, or books for high school teachers and science fair projects. Some come from reading newspaper accounts of chemical accidents.

One of the most difficult things in the beginning was dealing with difficult trainees. I had one experience where a trainee challenged me about some material in one chapter, and just went on and on. Most trainees are not like that, and most other people in the room don't like it when someone does that, either. Answer the person's question the best you can, offer to talk more on the break, and move on to the topic matter. I think it helps me that I'm 5'1" and female, and I'm probably very non-threatening to most people. I can say things in a kidding way, and people may not be as confrontational with me as with some other trainers.

Being seen as a peer really helps, and if people can view trainers more as a fellow worker that helps. If I know the group will be wearing jeans, and we're training in their workplace, I'll wear jeans, too. If I show up in a suit and heels, they're going to think "How long has she been in a paper mill with those heels on?" Talking to trainees on breaks, just everyday conversation, helps them be more attentive when you get back in the classroom. If someone else is being inattentive, they tend to calm the other person down so they're not disruptive. Peer enforcement helps you in the classroom.

Appendices

These diagrams and worksheets are used in participatory exercises, ranging from group problem-solving exercises to full-blown field scenarios. They are all from emergency response courses of varying levels, and are provided as examples of the thought processes and level of detail that work well in participatory training. They can serve as guidelines to developing similar exercises for any topic.

Each exercise has required prerequisites; for example, the Personal Protective Equipment exercise is preceded by classroom demonstration of all the information trainees need to perform the exercise. In a multiple-skill exercise such as the XYZ Products scenario, which serves as planning for an actual simulated response, trainees put into practice everything they have learned during the entire week's course. None of these exercises is designed to substitute for classroom, and possibly prior field, training.

PERSONAL PROTECTIVE EQUIPMENT WORKSHOP
Instruction Set

This exercise is intended as an alternative to a lecture format for reviewing and reinforcing basic knowledge related to PPE. It is set up to roughly follow the order of topics as presented in the PPE module currently used by CLEAR. This exercise should be followed by the exercise in levels of protection (attached).

This is a practical exercise and requires that an exercise area be set up within the classroom or some other room. Each exercise station is designated by a station sign and equipped with the items listed below. Students rotate from station to station and perform the tasks as required to complete the answer sheet (attached). After completion of the exercise, the instructor leads the class in a review of each station. In so doing PPE-related learning objectives can be achieved.

The stations should be supplied with the following items:

Station 1:   1 half-mask twin-cartridge air-purifying respirator, equipped with organic vapor cartridges
1 full-facepiece canister-type air-purifying respirator equipped with an organic vapor canister

Station 2:   1 fullface supplied air (or airline) respirator, with or without an escape air supply (note that the answer to part D varies according to the presence or absence of this feature)

Station 3:   1 fullface pressure-demand self-contained breathing apparatus

Station 4:   1 self-contained breathing apparatus which has several obvious defects such as the following:
- air cylinder well past date for hydrostatic pressure retest
- air supply low (e.g. approximately 1000 psig.)
- clouded, scratched, or crazed facepiece visor
- twisted, improperly attached harness
- damaged or missing exhalation valve
- etc.

Station 5: Several copies of drawings labeled A, B, and C depicting degradation, penetration, and permeation of chemical protective clothing (attached).

Station 6: Several copies of the *Quick Selection Guide to Chemical Protective Clothing* by Forsberg and Mansdorf (available through John Wiley and Sons) showing qualitative data on chemical protective materials.
Vendor information from Life-Guard(attached) showing quantitative data on chemical protective materials.

Station 7: 1 totally-encapsulating chemical protective suit
1 chemical-protective splash suit

Station 8: 1 totally-encapsulating chemical-protective suit which has various defects such as the following:
- damaged seam(s)
- missing or damaged exhalation valves
- abrasion or other physical damage to suit material (e.g. a torn glove)
- visible evidence of previous chemical exposure in one or two small areas (e.g. a butyl suit splashed with kerosene into which a small amount of used motor oil has been dissolved, then removed leaving an area of discoloration)
- components not properly attached (e.g. hardhat-to-suit attachments which are broken or not fully connected) (e.g. boots or gloves which are not properly attached to suit)
- physical abrasion to the right knee
- chemical degradation to the left knee (which can be produced by exposing the material to some product such as paint stripper to produce degradation, then decontaminating the garment.
NOTE: It may be advisable to label the knees of the pants to prevent confusion about which is which.

Station 9: Various accessory items for PPE ensembles or other types of safety equipment, such as
- cooling vest
- duct tape
- two-way radio (respirator compatible)
- harness and lifeline
- flash cover

For this workshop to be fully effective, it must be followed by thorough class discussion. The instructor should guide discussion into relevant topics related to PPE. This will insure that learning objectives are achieved. For example, in reviewing station 6, instead of merely pointing out the correct answers, the instructor should lead the class in a discussion of basic concepts of CPC selection. This should include discussion of the potential problems involved in CPC selection.

The answer sheet given to trainees follows, starting on page 126. Questions and instructions are on the answer sheet.

STATION 6, PART B

Chemical Permeation Test Results (ASTM F739 Method)

Liquid	LIFE-GUARD BUTYL MATERIAL Average* Breakthrough Time—minutes	Average** Perm. Rate (µg/min·cm²)	LIFE-GUARD NEOPRENE MATERIAL Average* Breakthrough Time—minutes	Average** Perm. Rate (µg/min·cm²)	LIFE-GUARD VITON/CHLOROBUTYL Average* Breakthrough Time—minutes	Average** Perm. Rate (µg/min·cm²)
Acetone	125	1	18	34	90	1
Acetonitrile	120	1	42	9	120	1
Carbon Disulfide	2	380	5	380	>480	0
Dichloromethane	4	583	6	1633	16	101
Diethylamine	3	527	16	567	13	143
Dimethylformamide	>480	0	60	107	>480	0
Ethyl Acetate	28	19	17	213	49	16
n-Hexane	4	487	20	80	>480	0
Methanol	303	1	210	3	392	1
Nitrobenzene	>480	0	45	49	>480	0
Sodium Hydroxide	>480	0	>480	0	>480	0
Sulfuric Acid	>480	0	>480	0	>480	0
Tetrachloroethylene	2	10	17	967	>480	0
Tetrahydrofuran	9	333	11	537	22	103
Toluene	6	770	12	920	>480	0

* Average time in minutes between contact of chemical on outside of material surface and detection of chemical on inside surface.
** Average rate at which the chemical permeates the material.
Tests performed by Radian Corporation. Before deciding if either of these materials will work in a particular situation, a swatch of the material should be tested against the chemical hazard.
There are uses and chemicals for which these accessories are unsuitable. It is the responsibility of the user to verify that these items are appropriate for the intended use and meet all health standards.

STATION 9, PART A

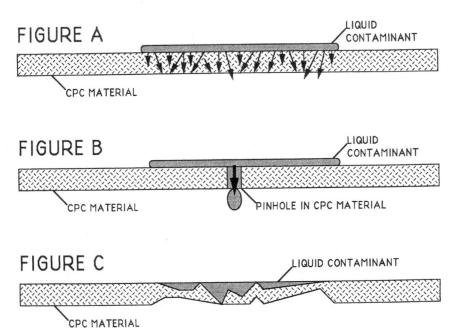

FIGURE A

LIQUID CONTAMINANT

CPC MATERIAL

FIGURE B

LIQUID CONTAMINANT

PINHOLE IN CPC MATERIAL

CPC MATERIAL

FIGURE C

LIQUID CONTAMINANT

CPC MATERIAL

EXERCISE IN LEVELS OF PROTECTION

In preparation for this exercise, the class will be divided into four groups. Each group will be assigned one of the four EPA-recommended levels of protection. Each group will then select the items needed to create a PPE ensemble representing the assigned level of protection. The required items will be selected from a variety of respirators, chemical protective garments, and accessory items provided for this exercise.

After all groups have completed the assignment, the class will review each level of protection. Each group should be able to explain the characteristics of their ensemble, including distinguishing features, advantages, and limitations as compared to other levels of protection. Each group should also be able to describe a task and work environment for which their ensemble would be appropriate.

The following items will be provided:

SCBAs
APRS
Totally-encapsulating chemical-protective suits
Chemical-protective splash suits
Coveralls
Surgical gloves
Chemical-protective gloves
Leather work gloves
Chemical-protective boots
Overboots
Boot covers
Leather safety boots
Hard hats
Safety glasses
Duct tape

PERSONAL PROTECTIVE EQUIPMENT WORKSHOP
Answer Sheet

STATION 1: RESPIRATORY PROTECTIVE EQUIPMENT

(A) Observe the two respirators on display at this station. These are examples of what type of respirator?

_____

(B) List any advantages you can think of which this type of respirator may offer, as opposed to other types of respirators.

_____
_____
_____

(C) List any disadvantages of this type of respirator.

_____
_____
_____

(D) Both the respirators at this station are equipped to protect the wearer against organic vapors. Which of the two should offer protection against the highest concentration of organic contaminants and why?

_____
_____
_____

STATION 2: RESPIRATORY PROTECTIVE EQUIPMENT

(A) What type of respirator is on display at this station?

_____

(B) List any advantages you can think of which this type of respirator may offer as opposed to other types.

_____
_____
_____

(C) List any disadvantages of this type of respirator.

_____
_____

(D) Is this item suitable for use during IDLH entries?

_____

Why?

_____

STATION 3: RESPIRATORY PROTECTIVE EQUIPMENT

(A) What type of respirator is on display at this station?

_____

(B) List any advantages you can think of which this type of respirator may offer as opposed to other types.

_____
_____
_____

(C) List any disadvantages of this type of respirator.

_____
_____
_____

STATION 4: SCBA INSPECTION

(A) Visually inspect the SCBA carefully and note any flaws which should be corrected before the unit is used in a contaminated area. List three flaws in the spaces provided below.

Flaws: _____

_____
_____

(B) Based on the SCBA air supply as indicated by the cylinder gauge reading, provide the following information.

Approximate Cylinder Gauge Reading (psi):_____
Estimated Total Breathing Time Remaining:_____
    (HINT: Allow I minute per 100 psi gauge pressure.)
Estimate of Breathing Time Before Alarm Sounds:_____
    (HINT: Alarm should go off at about 550 psi.)

2

STATION 5: CHEMICAL ATTACKS ON CPC

The figures displayed at this station represent three different ways in
which chemicals may attack or bypass chemical protective garments.
Match the letters from the drawings with the three types of chemical
attack listed below.

_____ permeation
_____ penetration
_____ degradation

STATION 6: CPC SELECTION

Use the provided information sources on chemical protective materials to
perform the tasks and answer the questions listed below.

(A) Use the *Quick Selection Guide to Chemical Protective Clothing* to
select the CPC material which should, according to the data provided,
offer the best resistance for the chemicals listed below. To save
time, the page number is provided for each chemical.

CHECK RECOMMENDED MATERIALS ✓

CHEMICAL HAZARDS	Butyl Rubber	Neoprene	Viton
Acetaldehyde			
Toluene			
Guthion			

(B) Based on the information gathered in part A, what conclusion can you
draw about the importance of identifying chemical hazards before
selecting CPC?

_____

_____

_____

3

(C) Using the table provided by Life-Guard, Inc., note the data supplied for toluene. Which of the three materials listed would offer the best resistance to toluene?

_____

How can you tell?

_____

(D) How does the type of data provided by the Life-Guard, Inc. table differ from that provided by the *Quick Selection Guide* ?

_____
_____
_____

STATION 7: CHEMICAL PROTECTIVE GARMENTS

(A) Identify the types of chemical protective clothing on display at this station.

(1) _____
(2) _____

(B) How do the two types of garments differ with regard to chemical protection?

_____
_____
_____
_____

(C) What degree of thermal protection (for example, protection from flash fire) are the garments on display at this station intended to provide?

_____
_____
_____
_____

4

STATION 8: TECP SUIT INSPECTION

(A) Carefully examine the TECP suit on display at this station. Note any
     defects or damage to the suit. List at least three problems revealed
     by your inspection of the suit.

     _____

     _____

     _____

(B) Carefully observe the knees of the suit. Both knees have been
     damaged, but in completely different ways. Try to distinguish
     between the type of damage to the right knee as opposed to the
     damage to the left knee.

     _____

     _____

     _____

     _____

STATION 9: OTHER SAFETY EQUIPMENT

Identify the accessory items on display at this station and briefly
describe how they might be used as part of a PPE ensemble.

_____

_____

_____

_____

5

CONTAINER IDENTIFICATION EXERCISE

Move around the room until you have been to each of the 8 stations. At each station, look at the exhibit and fill out this answer sheet. YOU MAY USE YOUR BOOK AND ANY OTHER MATERIAL AS A REFERENCE.

You do not have to do the stations in order. Move to the nearest station that has room for you.

STATION 1: NONBULK CONTAINERS

For each of the nonbulk containers described, write the letter that matches the most likely contents.

CONTAINER	DESCRIPTION	LIKELY CONTENTS
Bags	Multiwall paper or plastic; paper lined with plastic Capacity: 1-100 pounds	_____
Bottles and Jars	Glass or plastic with stopper or lid Capacity: few ounces to several gallons	_____
Carboys	Glass or plastic, usually inside cushioned outer container; may have wire cage for ease of handling Capacity: up to 10 gallons	_____
Cylinders	Metal with valve connections and pressure safety features. Colors not standardized, and not reliable indicator of contents.	_____
Drums	Metal or plastic; metal with plastic liner. Head may open entirely, or have bung opening.	_____

STATION 2: DRUMS

Write down what type of chemicals you would expect to find in each of the containers:

1. _____

2. _____

3. _____

STATION 3: HIGHWAY CARGO TANKS

For each of the highway tankers shown in the drawing, write the letter of
the appropriate drawing in the space following the correct item below.

SPEC.	DESCRIPTION	LIKELY CONTENTS	DRAW-ING
1. DOT 406 (MC 306)	Oval cross section; single uninsulated aluminum shell; divided into 3 or more compartments with top mounted domes; discharge fittings at bottom center. Design Pressure: 3 psi	Flammable or combustible liquids and Class 6.1 poisons	_____
2. DOT 407 (MC 307)	Large diamater, circular cross section; usually insulated; support ribs visible if uninsulated (as in this case); Dome and fittings at top center in protective "splash box". Design Pressure: 25 to 50 psi	Low vapor pressure liquids: flammables and mild corrosives.	_____
3. DOT 412 (MC 312)	Small diameter, circular cross section; uninsulated, with exposed support ribs; top mounted valves and fittings, usually at rear, in protective "splash box". Design Pressure: 15 psi	Heavy corrosives	_____
4. MC 331	Large diameter, round cross section; rounded ends; bolted manway in rear; valving concealed between (or just forward of) rear wheels. Design Pressure: 100-500 psi	Liqufied compressed gases (e.g. LPG, and anhydrous ammonia)	_____
5. MC 338	Round cross section; constructed of inner and outer tanks, with vacuum between tanks; all valves and fittings contained in cabinet at rear. Design Pressure: 25 to 500 psi	Cryogenic liquids (e.g. helium oxygen, and hydrogen)	_____

STATION 4: TRUCK ACCIDENT

In the model situation, a truck has wrecked and is leaking. If the contents
are hazardous, they may endanger the nearby trailer park and creek, as well
as anyone who drives down the highway.

1. What hazard class is the material in the truck?

_____

2. What is the the product carried by this truck?

_____

3. List 3 clues that helped you determine what the product is.

_____

_____

_____

STATION 5: BULK CONTAINERS

For each of the bulk containers shown in the drawing, write the letter of
the appropriate drawing in the space following the correct item below.

TANK TYPE	DESCRIPTION	LIKELY CONTENTS	DRAWING
Floating Roof with Dome	Round with roof floating on surface of product, supported by pontoons or double deck. Added dome cover to protect internal floating roof from weight of rain or snow	Low flash point liquids with low vapor pressure, such as crude oil	_____
Covered Floating Roof	Round with roof floating on surface of product, with an additional pitched or conical roof. Vents at roof joint relieve pressure differential at loading or unloading	Low flash point liquids, with moderate vapor pressure	_____
Vertical Dome Roof	Vertical cylindrical tank, with dome shaped roof designed to fail in case of excessive pressure. Operating pressure of 2.5-15 psi. Found in many shapes and sizes at manufacuring plants	Flammable and combustible liquids, solvents fertilizers, almost any liquid	_____
High Pressure	Usually single shell with no insulation, painted to reflect heat from sunlight. Spherical, or horizontal cylinder with rounded ends	LPG, methane, anhydrous ammonia, high vapor-pressure flammable liquids	_____
Cryogenic Liquid	Tank within a tank, well insulated. Tall,cylindri-al. Often supported above ground on legs	Liquid oxygen, liquid nitrogen, liquid carbon dioxide	_____

6. PLACARD

Look at the placard at this station and answer the questions.

1. What is the hazard class name? _____

2. What is the the primary hazard of chemicals in this class if they spill and mix with incompatible materials?

_____

3. Give an example of a chemical in this class.

_____

7. PLACARD

1. What is the physical state of the chemical behind this placard?
Circle one.

   SOLID   LIQUID   GAS   DUST   POWDER

2. What is the primary hazard if it spills? _____

3. Name 2 chemical properties you should look up to determine the degree of hazard.

_____   _____

8. PLACARD

1. What is the hazard class of this chemical?

_____

2. How can you find out the name of the chemical?

_____

_____

3. What will it do to your body?_____

CONFINEMENT PRACTICE EXERCISE

Connect
garden hoses
from 2 outlets
using Y-connects
so that each
station has
a hose end
with
nozzle.

Erect barrier
signs to
protect
trainees.

Exercise takes
place at end
of dead end
street.

1
DRUM SPILL RECOVERY

8 bags mulch
1 roll plastic
scissors
2 shovels
wheelbarrow
pump & accessories
4 air bottles
drum of water
bung wrench

2
SECONDARY CONFINEMENT STRUCTURE

1 roll plastic
scissors
2 shovels
wheelbarrow

3
MANHOLE COVER

neoprene mat
1 roll plastic
scissors
2 shovels
wheelbarrow

4
STORM DRAIN INLET

sorbent boom
1 roll plastic
4 bags mulch
pipe plug unit
2 air bottles

CONTAINMENT PRACTICE EXERCISE

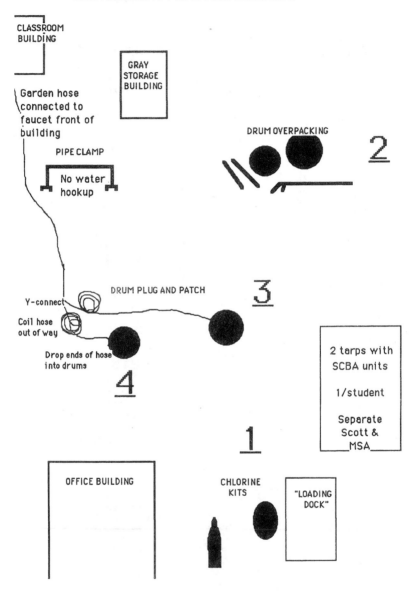

CONFINEMENT AND CONTAINMENT EXERCISE
WITH LEVEL A DRESSOUT

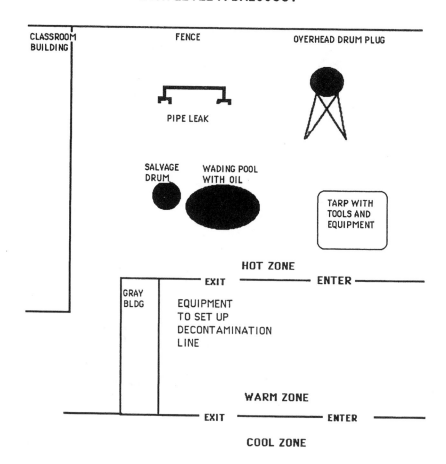

DRESSOUT AREA WITH ALL ASSOCIATED EQUIPMENT

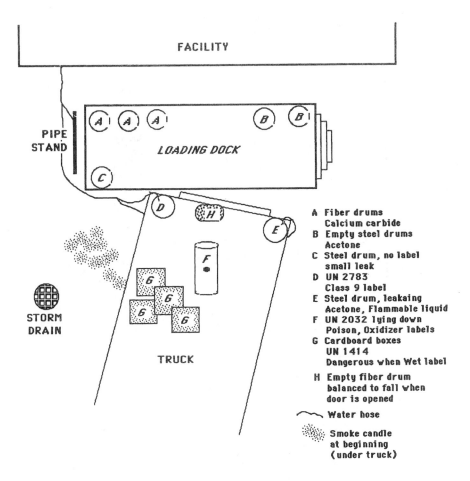

FACILITY

A Fiber drums
  Calcium carbide
B Empty steel drums
  Acetone
C Steel drum, no label
  small leak
D UN 2783
  Class 9 label
E Steel drum, leakaing
  Acetone, Flammable liquid
F UN 2032 lying down
  Poison, Oxidizer labels
G Cardboard boxes
  UN 1414
  Dangerous when Wet label

H Empty fiber drum
  balanced to fall when
  door is opened

〜 Water hose

Smoke candle
at beginning
(under truck)

AIR SURVEILLANCE EXERCISE

The following exercise sheets are used when we teach air surveillance. During this block of instruction, the students will do all three exercises as a way of breaking up the lecture and incorporating hands-on practice. We begin the block with a lecture/discussion of the basic principles of air surveillance and talk about the different kinds of air monitoring equipment. Then the class is broken into groups and each group is assigned an instrument and told to complete the Exercise 1. We have developed plain English instruction sheets for each instrument. Each group then teaches the rest of the class about their instrument based on the Exercise 1 Worksheet.

We then discuss instrument calibration and relative response. The same groups then complete Exercise 2 for their instrument, checking to see if their instrument is properly calibrated. This gives hands-on practice operating the instruments.

Finally, to wrap the class up, each group is given a gas bag that contains a contaminated atmosphere. They are told to take readings on their bag from all of the instruments and colorimetric tubes that are available. Based on their results they try to tell what chemical and what concentration they have. While giving practice, this exercise also reinforces the importance of proper training to proper use and interpretation of air surveillance equipment.

MEASURING CHEMICALS IN AIR
EXERCISE 1 WORKSHEET

Examine the instrument that was assigned to your group. Read the instructions for operating instrument that are provided. Answer the questions below and prepare to tell the rest of the class about the instrument. Start the instrument according to the instructions and become familiar with how it operates. If you have questions, ask an instructor.

What kind of chemicals or hazards does your instrument detect?

How do you turn your instrument and get readings?

How does the instrument show the results? What units (ppm, % LEL, % in air) are the results displayed in?

MEASURING CHEMICALS IN AIR
EXERCISE 2 WORKSHEET

Follow the instructions in the manual to turn on your instrument and allow it to warm up. With the equipment and calibration gas cylinder provided, check to see if your instrument is properly calibrated. Do not make any adjustments to the instrument unless instructed to by the instructor. Record below the expected and actual readings for each sensor in your instrument. If you are unsure about what the expected readings are, ask an instructor.

SENSOR	EXPECTED READING (from cylinder)	ACTUAL READING (from instrument)

Is your instrument properly calibrated?

MEASURING CHEMICALS IN AIR
EXERCISE 3 WORKSHEET

Take the gas bag your group has been assigned to different instrument and
detector tube stations and obtain readings. If you are unsure about how
the instrument works, an instructor will assist you. Record your readings
from each sensor in each instrument below. Make any calculations that
may be necessary based on relative response factors.

INSTRUMENT - SENSOR	READING	RELATIVE RESPONSE	ACTUAL CONCENTRAION
DETECTOR TUBE - BRAND	CHEMICAL	# OF PUMP STROKES	CONCENTRATION

Three "Ingredients" of Fire

What are the three ingredients of fire? Label the Fire Triangle below.

List examples of each of the following that are present in your workplace.

Fuel Sources:

_____

_____

_____

_____

Oxygen Sources:

_____

_____

Energy (Ignition) Sources:

Electrical:_____

_____

Chemical: _____

_____

Thermal:_____

_____

Mechanical: _____

_____

Radiation:_____

_____

Based on your answers on the previous page, how might fire be prevented in your workplace.

_____
_____
_____
_____
_____
_____
_____
_____
_____
_____
_____

Leading Causes of Industrial Fires

Put the following in order from most common to least common cause of Industrial Fires.

_____ Cutting and Welding

_____ Overheated Materials

_____ Smoking (cigarettes, etc.)

_____ Sparks

_____ Chemical Action

_____ Electrical (wiring and motors, poor
                   maintenance of equipment)

_____ Hot Surfaces, Burner Flames, Combustion
                   Sparks

_____ Spontaneous Ignition

_____ Arson

_____ Friction (bearings, jammed machines, etc.)

Exercise in Fire Extinguisher Selection

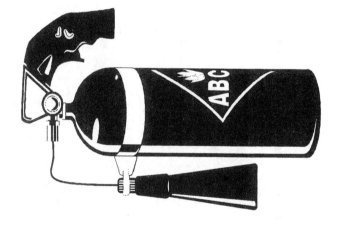

Description Of Fire	Classification (A,B,C,or D)	Extinguisher Types (Check all that apply)					
		Pressurized Water A	Aqueous Film Forming Foam A,B	Dry Chemical B,C	Multi-Purpose Dry Chemical A,B,C	Carbon Dioxide B,C	Combustible Metals D
Paper in Trash Can							
Fire in Engine Compartment Of Truck							
Over-loaded Extension Cord							
Hexane Spill							
Pile of Scrap Wood & Trash							
Computer							
Pile of Mg Shavings							

ASSESSING THE HAZARDS OF CHEMICALS
EXERCISE

On the following page is a worksheet for determining chemical hazards. Look up a chemical that you use in your workplace or at home, or that is used in the facility where you work. Fill out the worksheet and use this information to determine the fire, explosion, reactivity, and exposure hazards of each chemical based on the properties you find. Use any or all of the following references:

Material Safety Data Sheets
NIOSH Pocket Guide to Chemical Hazards
Dangerous Properties of Industrial Materials (Sax)
Association of American Railroad Guides
Merck Index
New Jersey Fact Sheets
Frank Fire's Chemical Data Notebook

CHEMICAL HAZARD WORKSHEET

Name of Chemical _____ DOT Id. # _____

Physical state at normal temperature (solid, liquid, gas)_____

Fire, explosion, and reactivity hazards:

Flash point, in ^{0}F _____   Is this chemical flammable?

Is flash point below normal   If so, under what conditions?
working temperature? _____

LEL _____   Does the chemical have a wide

UEL_____   flammable range?

Is this chemical an oxidizer that may
react with other flammables in the
area to start a fire?_____

Does this chemical react strongly with:

Air? _____   Other chemicals?_____

Water?_____   What other chemicals?

Is this chemical stable?_____

Are harmful products of reaction listed?

What are they?

Vapor pressure is   Is it likely to explode its container
_____ mmHg   if the temperature goes up? _____

Dispersion on a site:

Solubility_____

Specific Gravity_____

Vapor Pressure_____

Vapor Density_____

Boiling Point_____

Based on these properties, will
A large vapor cloud be produced?

Where will the chemical go in water?

Will it float away or sink?

Where will it go on land?

Exposure hazards:

If there is a spill while you are working, where will the exposure hazard be?
(In breathing air, by skin contact, in water bodies?)

If the chemical is in the soil on the site, what kinds of activities will lead to
possible worker exposure?

If the chemical has been in the water on the site for a long time, is it still
likely to be there?

If so, what kinds of work will lead to possible worker exposure?

HEALTH EFFECTS OF CHEMICALS
WORKSHEET

Use your NIOSH Pocket Guide, MSDS, and/or any other reference materials available to you to fill out this worksheet. Not all chemicals have all this information available for them. Complete what you can and write NA in blanks where you do not have the information.

Name of Chemical _____

8 – Hour TWA   OSHA (PEL) _____
                           NIOSH (REL)_____
                           ACGIH (TLV)_____

Short Term Limit (STEL) (If it has one) _____

Ceiling Limit (C) (If it has one) _____

Route(s) of Entry:

Symptoms of Exposure:

Target Organs:

Do you think this chemical can cause you serious harm or disease?

If so, what kind of harm or disease?

CHEMICAL HAZARD WORKSHEET

Use your NIOSH Pocket Guide and the materials on the reference table in the back of the room to fill out the worksheet.

Name of Chemical_____

Exposure limits NIOSH_____ OSHA_____
Short term limit (if it has one)_____
IDLH_____

Route(s) of entry:

Symptoms of exposure:

Target Organs:

Physical description of the chemical:

Flammability: Flash Point_____
                 LEL   _____
                 UEL   _____

Vapor Pressure_____

Molecular Weight_____

Vapor Density (RGasD)_____

Solubility_____ Specific Gravity _____

Incompatibilities and Reactivities:

Fire and Health Hazard Information Worksheet

NAME OF CHEMICAL	Route Of Entry	What does label look like?	Target Organs	8 Hr. TWA Exposure Limit	Symptoms and Illness	Flash Point	LEL And UEL	Vapor Pressure	Vapor Density

RESOURCE MANAGEMENT USING THE INCIDENT COMMAND SYSTEM

Part 1: Exercise in Resource Needs Assessment

This exercise is based directly on the information you produced in completing the Hazard and Risk Assessment Exercise. Your team will be required to fully assess the resources (equipment, tools, supplies, and personnel) required to carry out the response options you identified in the previous exercise.

Work together with the other members of your group to fill out the resource management worksheet(s) on the following page(s), indicating the resources required for achieving your objective(s). List your response options in order of decreasing preference as determined in the Hazard Assessment Exercise. Use the worksheet to assess the resource needs for each option. For an example, see the sample worksheet on the following page.

Each group should record its conclusions on a single worksheet or set of worksheets to be used later in planning. Also, each group should elect a representative who will act as spokesperson in presenting the group's findings to the class.

Part 2: Managing Human Resources: Review of Incident Command System Basics

In order to do the response options listed in the first column of the worksheet, the activities of personnel listed in the fifth column must be carefully controlled. You may recall that the HAZWOPER standard requires that we use the Incident Command System (ICS) for this.

We will review the basics of ICS by viewing a video: *The Incident Command System for Hazardous Materials Incidents: Overview of the Incident Command System* by Action Training Systems, Inc.

As you watch the video, think about how the ICS could be best organized to carry out the response options identified by your group. In the planning session which follows, the class will be organized into a command structure for responding to an incident at the XYZ products site.

WORKSHEET FOR RESOURCE MANAGEMENT

OBJECTIVE: Prevent spilled liquid from entering waterway

Response Option	Activities Involved	Resource Requirements			Other Considerations or Comments
		Equipment	Supplies	Personnel	
E.G. Build dike across ditch to confine spilled liquid	Excavate approximately 7 cubic yards of soil from hillside and place across ditch. Line with plastic.	1 backhoe/ loader 2 shovels	plastic	1 equipment operator 2 laborers	Dump truck and operator may be required depending on distance soil must be transported.

WORKSHEET FOR RESOURCE MANAGEMENT

OBJECTIVE:

Response Option	Activities Involved	Resource Requirements			Other Considerations or Comments
		Equipment	Supplies	Personnel	

XYZ PRODUCTS COMPANY
RAIL YARD
CHEMICAL INVENTORY

NAME	AMOUNT(MAX)	CONTAINER TYPE/CAPY.
Acetic Acid, Glacial	2 Rail Cars	DOT 111A100W2 (13,600 gal)
Ammonia, Anhydrous	2 Rail Cars	DOT 112S400W (33,500 gal)
Benzoyl Peroxide (paste with 40% H20)	1 Rail Car	DOT 111A60W1 (14,150 gal)
Chloroform	2 Rail Cars	DOT 111A100W1 (20,000 gal)
Gasoline	2 Rail Cars	DOT 111A100W1 (20,000 gal)
LPG	2 Rail Cars	DOT 105J300W (24,750 gal)
Nitrogen	2 Rail Cars	DOT 113C120W (25,500 gal)
Propylene Glycol	1 Rail Car	DOT 111A60W1 (23,150 gal)

XYZ PRODUCTS COMMUNITY MAP

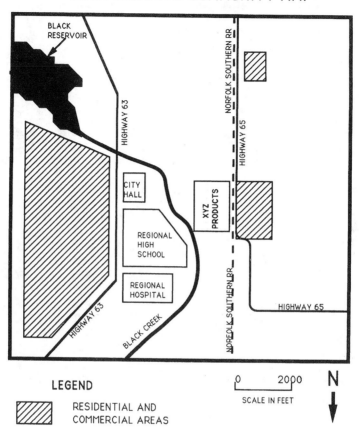

LEGEND

RESIDENTIAL AND
COMMERCIAL AREAS

0   2000
SCALE IN FEET

N

XYZ PRODUCTS LOCATION MAP

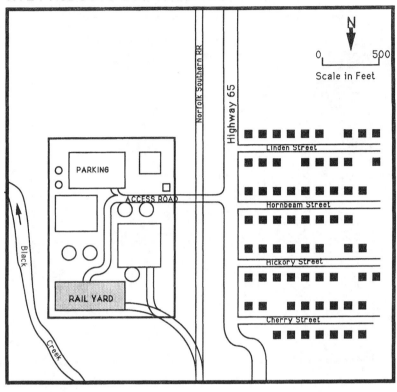

LEGEND

———————	XYZ Products Property Fence
▪	Occupied Dwelling
▨	Rail Yard (see attached enlarged map)

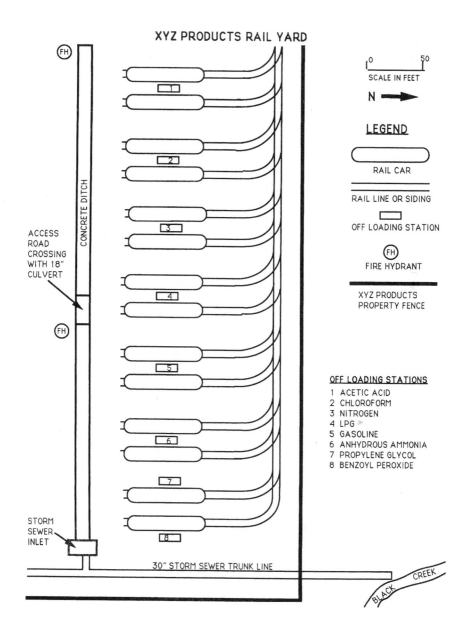

XYZ PRODUCTS RAIL YARD

SCALE IN FEET

N →

LEGEND

RAIL CAR

RAIL LINE OR SIDING

OFF LOADING STATION

FIRE HYDRANT

XYZ PRODUCTS
PROPERTY FENCE

OFF LOADING STATIONS
1 ACETIC ACID
2 CHLOROFORM
3 NITROGEN
4 LPG
5 GASOLINE
6 ANHYDROUS AMMONIA
7 PROPYLENE GLYCOL
8 BENZOYL PEROXIDE

ACCESS
ROAD
CROSSING
WITH 18"
CULVERT

CONCRETE DITCH

STORM
SEWER
INLET

30" STORM SEWER TRUNK LINE

BLACK CREEK

ACETIC ACID, GLACIAL (liquid)

Flash point
Flammable limits
Vapor pressure
Vapor density
Solubility
Ionization potential
Reactivity and incompatibility
Routes of exposure
Exposure limits
Target organs
Symptoms of exposure
Products of overheating or combustion
Other relevant information

ANHYDROUS AMMONIA (liquefied compressed gas)

Flash point	
Flammable limits	
Vapor pressure	
Vapor density	
Solubility	
Ionization potential	
Reactivity and incompatibility	
Routes of exposure	
Exposure limits	
Target organs	
Symptoms of exposure	
Products of overheating or combustion	
Other relevant information	

BENZOYL PEROXIDE (solid – shipped as a paste)

Flash point
Flammable limits
Vapor pressure
Vapor density
Solubility
Ionization potential
Reactivity and incompatibility
Routes of exposure
Exposure limits
Target organs
Symptoms of exposure
Products of overheating or combustion
Other relevant information

CHLOROFORM (liquid)

Flash point	
Flammable limits	
Vapor pressure	
Vapor density	
Solubility	
Ionization potential	
Reactivity and incompatibility	
Routes of exposure	
Exposure limits	
Target organs	
Symptoms of exposure	
Products of overheating or combustion	
Other relevant information	

GASOLINE (liquid)

Flash point
Flammable limits
Vapor pressure
Vapor density
Solubility
Ionization potential
Reactivity and incompatibility
Routes of exposure
Exposure limits
Target organs
Symptoms of exposure
Products of overheating or combustion
Other relevant information

LIQUIFIED PETROLEUM GAS

Flash point	
Flammable limits	
Vapor pressure	
Vapor density	
Solubility	
Ionization potential	
Reactivity and incompatibility	
Routes of exposure	
Exposure limits	
Target organs	
Symptoms of exposure	
Products of overheating or combustion	
Other relevant information	

NITROGEN (cryogenic liquid)

Flash point
Flammable limits
Vapor pressure
Vapor density
Solubility
Ionization potential
Reactivity and incompatibility
Routes of exposure
Exposure limits
Target organs
Symptoms of exposure
Products of overheating or combustion
Other relevant information

PROPYLENE GLYCOL (liquid)

Flash point	
Flammable limits	
Vapor pressure	
Vapor density	
Solubility	
Ionization potential	
Reactivity and incompatibility	
Routes of exposure	
Exposure limits	
Target organs	
Symptoms of exposure	
Products of overheating or combustion	
Other relevant information	

If this chemical is released, what are the possible bad outcomes? In other words, use what you have found out about the chemical to ESTIMATE LIKELY HARM WITHOUT INTERVENTION.

Harm to the physical facility:

Harm to responders:

Harm to workers in the immediate area:

Harm to members of the surrounding community:

Harm to the environment:

Use the back of the sheet if you need more space

CHOOSING OBJECTIVES AND IDENTIFYING OPTIONS

What actions can you as responders take in order to prevent the harm likely to result from an incident involving this chemical? To answer this question, use steps 3 and four of the DECIDE process.

In **step 3**, your group will **choose objectives** for preventing or minimizing the harm you have identified as likely to result from a release of your chemical.

After listing all your response objectives, go back and **identify options** available for achieving each objective. This is **step 4** of the DECIDE process.

STEP 3: CHOOSE RESPONSE   *STEP 4: IDENTIFY OPTIONS FOR*
OBJECTIVES   *ACHIEVING OBJECTIVES*

RESPONSE OBJECTIVE:

OPTIONS FOR ACHIEVING THIS OBJECTIVE:

RESPONSE OBJECTIVE:

OPTIONS FOR ACHIEVING THIS OBJECTIVE:

**STEP 3: CHOOSE RESPONSE
OBJECTIVES**

**STEP 4: IDENTIFY OPTIONS FOR
ACHIEVING OBJECTIVES**

RESPONSE OBJECTIVE:

OPTIONS FOR ACHIEVING THIS OBJECTIVE:

RESPONSE OBJECTIVE:

OPTIONS FOR ACHIEVING THIS OBJECTIVE:

RESPONSE OBJECTIVE:

OPTIONS FOR ACHIEVING THIS OBJECTIVE:

CONSIDERATIONS IN IDENTIFYING RESPONSE OPTIONS

PERSONAL PROTECTIVE EQUIPMENT SELECTION

Respiratory protection for potential air contaminant concentrations

Concentration (list units)	Respiratory protection

Level of protection for defensive actions (circle one)

A   B   C   D

Level of protection for approaching the leak and stopping it (circle one)

A   B   C   D

Glove and clothing fabric selection

SORBENT COMPATIBILITY

Compatible materials	Incompatible materials

COMPATIBILITY OF PLUGS, PATCHES, GASKETS, HOSES

Compatible materials	Incompatible materials

DILUTION AND NEUTRALIZATION

Will you plan to dilute or neutralize the chemical if it is released?

Methods and considerations:

VAPOR SUPPRESSION CONSIDERATIONS

Will hazardous vapors likely be present in large quantities?

If so, should you plan to suppress the vapors? Why or why not?

Methods:

PUBLIC PROTECTION

Using the information available to you, plan to evacuate or protect in place members of the community surrounding your facility. Organize your plan, and give reasons for your choices. Also consider the resources necessary to do either or both.

LESSON PLAN

HAZARDOUS MATERIALS UPDATE

HAZARDOUS MATERIALS INCIDENT ASSESSMENT

TOTAL TIME: 1 Hr. 30 Min.

STUDENT MATERIALS: Handouts with instructions, scenario descriptions, and worksheets (the numbered pages and illustrations that follow)

TRAINING MATERIALS NEEDED:

–Overhead Projector
-Slide Projector
-Flip Chart and Markers
-Transparencies of drawings for scenarios 1,3,&6
-Models for scenarios 2 and 4
-Photo or slide for scenario 5

TRAINEE REFERENCES:

-1996 DOT Emergency Response Guidebook
-NIOSH Pocket Guide to Chemical Hazards
- MSDS's
-Commonly used hazmat references, such as:
    CHRIS Manual
    AAR's Emergency Action Guides
    Frank Fire's Chemical Data Notebook
-GATX Tank Car Manual (for Scenario 4)
-Chlorine Institute Booklets (for Scenario 5),
    including:
    -"The Chlorine Manual"
    -"Emergency Kit "B" Instructions"
-Other items as needed or requested

INSTRUCTOR BACKGROUND REFERENCE

"Emergency Responder Training Manual for the Hazardous Materials Technician" by UAB/CLEAR, John Wiley & Sons, New York, Chapter 5

OBJECTIVES

After completing this module, the learner will be able to:
- Use Ludwig Benner's DECIDE process in assessing a hazardous materials incident, including:
  -**D**etection of the presence of hazardous materials
  -**E**stimation of likely harm without intervention
  -**C**hoice of response objectives
  -**I**dentification of options for achieving response objectives
  -**D**etermination of which option is best
  -**E**valuation of progress as response option is carried out
- Use commonly available references in gathering information for use in incident assessment

TEACHING OUTLINE

INTRODUCTION
(Time: 10 Min.)
- Focus on the importance of using structured decision-making guidelines, such as the DECIDE process, during assessment and response.
  - Refer to the steps of the DECIDE process using key words written on a flip chart with the first letter of each step printed boldly in contrasting colors and spelling the word "DECIDE" when read vertically.

SMALL GROUP WORK IN INCIDENT ASSESSMENT
(Time: 35 Min.)
PROCEDURE:
1. Divide the entire class into 6 teams. Be sure that the knowledge and expertise of the class are evenly divided among the teams and not concentrated into one or two teams. Assign one scenario to each team.

2. Provide exercise handouts to trainees and review instructions on page one of handout.
   - Put special emphasis on the tasks which each team is to complete for their scenario.
   - Be sure that the class is aware of the worksheets located at the end of the instructions and how they can be useful in completing the assignment.

3. After teams have begun work, circulate constantly from group to group and work with each as needed.
   - Inquire about how things are progressing and be sure that they fully understand their scenario.
   - Give each group a nudge in the right direction, as appropriate, if needed.

4. Provide each team any items which may assist the group representative in presenting the group's findings to the class. Examples of such items include:
   - Blank transparency sheets
   - Markers which can be used on the transparencies
   - Site map transparencies and slides of models and photos

PRESENTATION OF RESULTS
(Time: 40 Min.)
1. After all groups have completed the assigned tasks, orchestrate the presentations of each group's findings. Assist each presenter as needed (e.g. in operating AV equipment, etc.).

2. After each presentation:
   - Request additional comments from the other members of the group.
   - Request questions or comments from the class at large.
   - Summarize the scenario and the group's findings in the fashion of a critique, with a special emphasis on the good points. Call into question any findings that are unclear or seem incorrect. Ask for clarification from the group and discussion from the rest of the class in an attempt to resolve these.

CONCLUSION
(Time: 5 Min.)
1. Offer a final chance for questions or comments from trainees.

2. Close with a reminder of the advantage of using decision making guidelines (such as the DECIDE process) in assessing an incident, as this helps avoid tunnel vision once adrenal starts to flow.

HAZARDOUS MATERIALS INCIDENT ASSESSMENT WORKSHOP

ASSIGNMENT

Work with the other members of your team to evaluate your assigned incident. Use the first five steps of Ludwig Benner's DECIDE process to assess the hazards involved and determine appropriate response actions. Be prepared to report your results to the rest of the class.

DETECT PRESENCE OF HAZARDOUS MATERIALS
Establish the specific identity of hazardous materials involved in your scenario.

ESTIMATE LIKELY HARM WITHOUT INTERVENTION
Establish a hazard profile for each hazardous material involved. This profile should include any toxicity, asphyxiation, etiological, thermal, explosion, radiation, reactivity, corrosivity, and/or environmental hazards.

Based on the results of the hazard profile and the specifics of the incident, determine the potential harm the incident represents to people (including responders), property, and the environment.

Ultimately, what will the likely adverse effects be if the incident is allowed to run its course?

CHOOSE RESPONSE OBJECTIVES
Establish reasonable goals for preventing or reducing likely harm resulting from the incident.

IDENTIFY OPTIONS FOR ACHIEVING RESPONSE OBJECTIVES
Determine the action pathways which can reasonably be taken in order to achieve the chosen response objectives.

DO BEST OPTION
This requires a weighing of the options identified in the previous step. Factors such as likelihood of success, resource requirements, and hazards to personnel must be compared for various options in order to determine which is best.

EVALUATE PROGRESS AS THE RESPONSE OPERATION UNFOLDS

1

SCENARIO 1 (see drawing)

A train derailment has occurred at a rail crossing on County Road 25. This is a rural location, with only three occupied dwellings within a one mile radius. The location is adjacent to a gully (see the drawing) which drains into the Cahaba River approximately 2000 feet to the west.

Nine rail cars have been damaged. With the help of the conductor and the train consist you are able to identify the contents of the rail cars involved in the derailment (as shown below the drawing).

The time is 1:00 p.m. Weather conditions are clear, with 60% humidity and a temperature of 42 degrees. Wind is from the west at 5 mph.

SCENARIO 2 (see model)

Assume the role of firefighter and hazmat team member for a municipal fire department. Your engine company is the first unit to respond to the report of a truck wreck involving a potential fire hazard. The time is 1:40 a.m.. Weather conditions are cloudy and humid, with a temperature of 70°F and wind from the west at 5 to 10 MPH, with gusts to 20 MPH.

As you approach the location of the accident, you see the scene represented by the model. A tank truck has been involved in a one vehicle, rollover-type accident. The tanker appears to have only minor damage. Liquid product is leaking from the cargo tank as indicated by the model, with an estimated 5 to 6 GPM rate of release. The driver reportedly sustained minor injuries and has been taken to the hospital by a passing motorist.

SCENARIO 3 (see drawing)

An accidental release of Chlorosulfonic acid occurred at XYZ Industries when a hose ruptured as a rail car was being unloaded (see the drawing) . Fuming of the spilled liquid product prevented XYZ personnel from approaching the rail car to close the discharge valve and stop the release.

The time is 11:30 a.m. Weather conditions are clear, with 80% humidity and a temperature of 80 degrees. Wind is from the west at 5 mph .

2

SCENARIO 4 (see model)

Assume the role of an industrial hazardous materials response team
member. You are investigating the report of an accident on a rail line
along the west boundary of your plant. The time is 3:20 p.m. and weather
conditions are clear, with a temperature of 43° F and a 10 MPH wind from
the north.

As you approach the scene, you notice a stationary train, the midsection of
which has obviously derailed. Your attention is immediately drawn to a
location at which flame and smoke are visible. You stop and examine this
location with binoculars. Three tank cars are involved in the fire. The
tank cars are situated as shown in the model.

SCENARIO 5 (see photo or slide)

Assume the role of a fire fighter/hazmat team member answering a call
involving a fire at a water treatment plant. The time is 11:00 a.m. and the
weather is partly cloudy with a temperature of 80° F and winds from the
east to southeast at 3 to 5 MPH.

Upon arrival, you are informed by a facility representative that a flash
fire occurred at the plant approximately 12 minutes before your arrival.
The fire was intense, but lasted only briefly before burning out. The fire
appeared to be due to a leaking section of gas line which has since been
isolated by closing valves. There appears to be no further hazard from
leaking gas at the moment. However, plant personnel who have just exited
the area where the fire occurred are reporting a pungent, choking,
bleach-like odor and irritation of their eyes and respiratory tracts.

The hazardous materials response unit is deployed and one team enters the
area in question for assessment. The assessment team reports that the
only chemical containers in the area affected by the fire are those shown
in the slide. The team also reports that one of these containers is leaking
from the fitting indicated by the arrow in the slide. The released product
appears as an amber liquid which boils vigorously to produce a greenish
yellow gas.

3

SCENARIO 6 (see drawing)

An accident involving a tractor trailer rig has occurred on Paper Mill Road
(see the drawing). The driver was able to walk away from the accident
and is still on the scene.

The time is 1:25 p.m. Weather conditions are overcast, with 95% humidity.
Intermittent rain is predicted for the afternoon. Temperature is 85
degrees and wind is from the south at 3 to 5 MPH.

4

SCENARIO 1

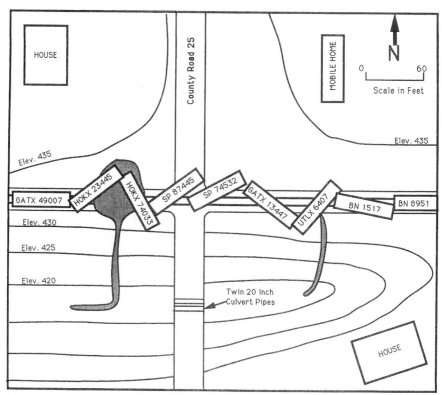

CAR	TYPE	CONTENTS	CONDITION
GATX 49007	Hopper	Corn	On rails, minor damage
HOKX 23445	Tank Car	Styrene Monomer, inhibited (UN 2055)	Derailed, upright, major damage, but no leakage apparent
HOKX 74033	Tank Car	Styrene Monomer, inhibited (UN 2055)	Derailed, upright, leaking from large puncture in head of car
SP 87445	Box Car	Auto Parts	Derailed, overturned, major damage
SP 74532	Box Car	Auto Parts	Derailed, overturned, moderate damage
GATX 13447	Tank Car	Corn Syrup	Overturned, moderate damage, no leaks
ULTX 6407	Tank Car	Sulfuric Acid (UN 1830)	Overturned, moderate damage, leaking from bolted manway fitting
BN 1517	Box Car	Garments	Derailed, upright, minor damage
BN 8951	Box Car	Garments	On rails, minor damage

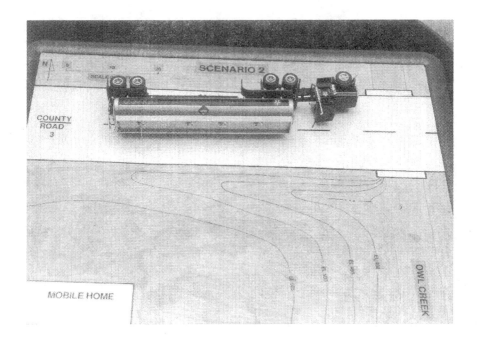

SCENARIO 3

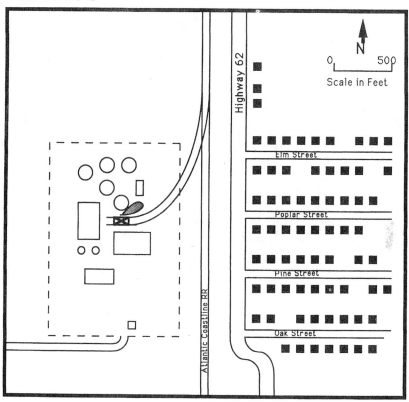

LEGEND

— — — — XYZ Industries Property Boundary

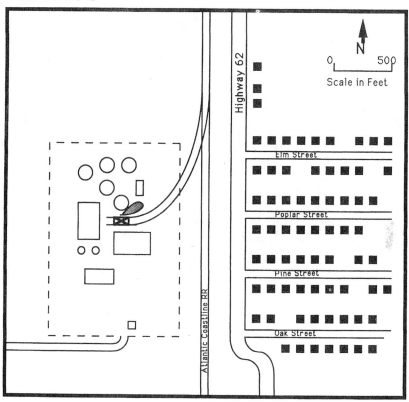 111A60W2 Rail Car with Corrosive and Poison placards and UN 1754 markings. Ruptured off-loading hose attached to car. Discharge valve is open.

Spilled liquid product

Occupied Dwelling

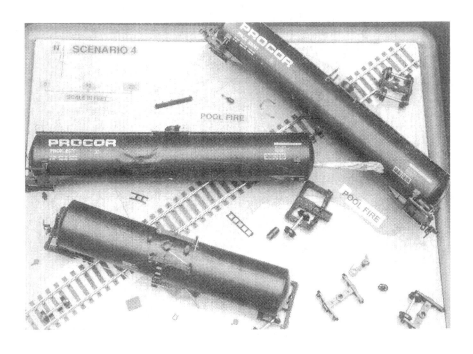

SCENARIO 6

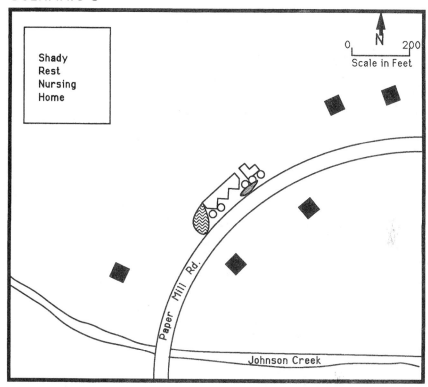

Shady
Rest
Nursing
Home

N

0   200
Scale in Feet

Paper Mill Rd.

Johnson Creek

LEGEND

Semi with pneumatic hopper trailer, overturned with major damage. Trailer markings: Un 1748 Remnant of yellow placard visible.

Diesel fuel leaking from saddle tanks of tractor.

White solid material spilled from trailer during rollover.

Occupied Dwelling

CHEMICAL HAZARD WORKSHEET

Name of chemical _____ DOT i.d. #_____

Physical state at normal temperature (solid, liquid, gas)_____

Fire, explosion, and reactivity hazards:

Flash point, in °F_____   Is this chemical flammable?
                                           If so, under what conditions?
Is the flash point below normal
working temperature (ambient)?

LEL_____   Does it have a wide flammable
                                           range?
UEL_____

Is it an oxidizer that may react
with other flammable materials
in the area to start a fire?

Does it react strongly with:
Air?_____   Other chemicals?_____

Water?_____   What other chemicals?

Is it stable?_____

Are harmful products of reaction listed?

What are they?

Solubility_____

Specific gravity_____

Vapor pressure_____

Vapor density_____

Boiling point_____

Is it likely to explode its container if the temperature goes up?

Will a large vapor cloud be produced?

Is there potential for vapors to seek out ignition sources?

Where will the chemical go in water? Will it float away or sink?

Where will it go on land?

Exposure hazards:

What are the chemical's routes of entry into the body?

Where will the main exposure hazard be? (in breathing air, skin contact, etc.)

What level of protection would be required if there has been a large release of this material?

chapter eight

Building hands-on training devices

One of the challenges you must address in going from classroom to field training is coming up with practice devices or "props" for hands-on training. As one example, let's consider emergency response training for hazardous materials technicians. According to the OSHA standard regulating Hazardous Waste Operations and Emergency Response (29 CFR 1910.120), hazardous materials technicians are responders who "approach the point of release in order to plug, patch, or otherwise stop the release of a hazardous substance." In other words, technician trainees need practice in using equipment and procedures for containing accidental releases or leaks of hazardous materials.

It is generally not practical to stage actual transportation accidents or intentionally damage containers or piping systems in industrial facilities for training purposes. In order to provide containment training, we must provide realistic simulated releases for trainees to repair (Figure 8.1).

In this chapter, we will look at basic considerations for creating training devices. These basic considerations will be illustrated with devices that have actually been created and used by the authors. Keep in mind that these devices are intended only as examples, and that the basic principles they illustrate can be applied in creating any type of device for any type of training.

I. Basic design considerations

Your imagination and sense of innovation are valuable tools for creating training equipment. The devices you produce can range from simple to complex in design, from mild to wild in concept, and from easy to difficult in terms of demands on trainees. It is good to keep a few basic principles in mind from the beginning.

In order to be useful, training devices must be:

Figure 8.1 Using water to simulate hazardous materials allows trainees to safely practice containing liquid releases.

- Available at an affordable price
- Appropriate for the type of training to be delivered
- Safe to create and use in training
- Interesting for trainees (and trainers also)

Let's apply these four principles to producing some devices to practice leak control for Hazmat technician training.

II. Some assembly required

All parts of the devices we will consider, except for the chlorine containers, are readily available through local sources. Sources include hardware stores, plumbing supply houses, and industrial supply companies. The fabrication and modification required to create the devices involves only minimal mechanical skills and commonly available tools. As a result, we were able to fabricate the items in house at minimal cost.

As in all work activities, follow good safety practices closely when creating training devices. For example, installing a fitting to pressurize a gas cylinder requires a hole to be drilled and tapped in the cylinder side wall. This procedure alone could involve hazards ranging from flying particles impacting the eyes while drilling to chemical exposure, fire, or explosion in the event the cylinder contains residual product. Such operations should be performed only by personnel familiar with the appropriate safety precautions. Remember: As a safety trainer, you will look especially foolish if you, or someone acting under your direction, are injured while creating a safety training device.

III. Leaking pipe stand

The pipe stand is a freestanding device assembled from water pipe (Figure 8.2). Build one using galvanized metal pipe and fittings for strength, or from

Figure 8.2   The pipe stand is used to simulate releases from pipes, valves, and fittings.

PVC tubing if you need a cheaper and lighter option. Install cam lever couplers, as shown in Figure 8.3, if you want ease of disassembly, transportation, and storage. Supply the pipe stand with water, using a garden hose or other suitable source, and you are almost ready to train. All you need now are leaks.

In creating training devices, use the learning objectives as a guide. Using containment training as an example, what types of repair operations do your trainees need to be able to perform? This will determine the types of leaks

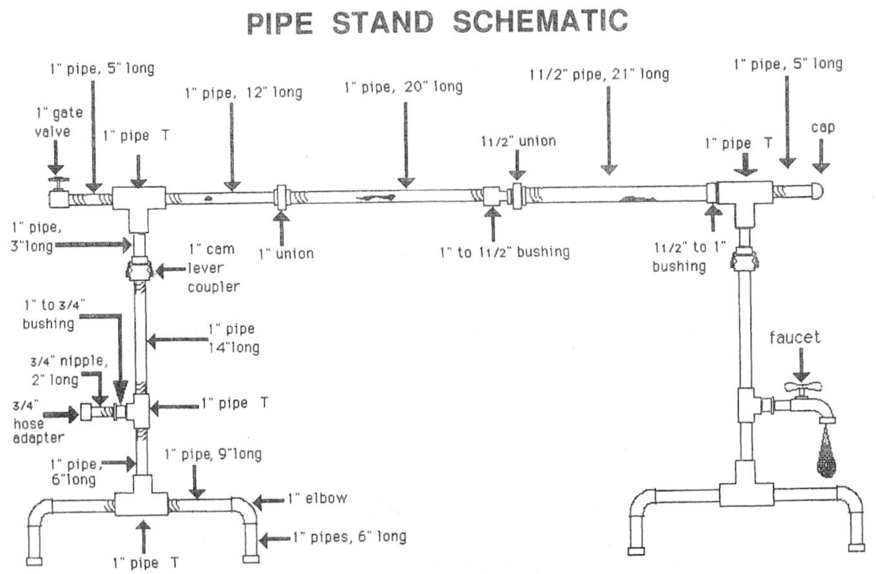

PIPE STAND SCHEMATIC

Figure 8.3   Pipe stands are cheap and easy to design and build. Skills required are minimal and parts needed are readily available. (Courtesy of D. Alan Veasey, from *Fire Engineering*, 149(11), 68, 1996.)

you provide (Figure 8.4). If you want your trainees to be able to use pipe repair clamps to stop leaks from holes in piping, then drill, cut, or grind holes of various sizes and shapes into the pipes. If your trainees need to be able to use wrenches, loosen connections or fittings on the pipe stand before beginning an exercise. For a simple but realistic point of release, simply leave a valve open.

Items like the pipe stand are useful for teaching good operating procedures. For example, show your trainees that by placing a container below the gate valve (Figure 8.3) and opening the valve fully before beginning repairs, they can reduce the pressure on the leaks to be repaired and confine most of the spillage occurring during the operation. This can reduce contamination of both the responders and the environment.

Versatility is one of the main advantages of devices like the pipe stand. Modify the size, type, and configuration of pipes, fittings, and the damage to them as needed. The only limits are the training objectives, trainee prerequisites, and your imagination.

IV. General containment training device

We built the general containment training device, also known as "Leak Monster," from the tank of a junked 30-gallon electric water heater. Use devices similar to the Leak Monster when you want to provide your trainees

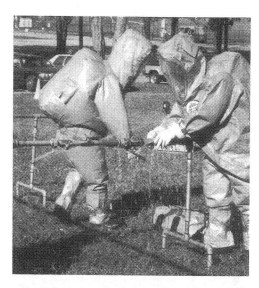

Figure 8.4   The pipe stand allows trainees to practice various containment procedures applicable to piping systems.

practice in a variety of general plug/patch or leak repair operations in a way they will enjoy.

To use this type of device, simply set it up and instruct your trainees to repair flowing leaks as they occur (Figure 8.5). Provide them with a tools-and-equipment set, but require them to select the items needed to complete repairs. As existing leaks are stopped, new leaks appear as the water level inside the tank rises.

Use a device like the Leak Monster to challenge your trainees to perform the following repairs.

1. Install a 1 1/2″ pipe repair clamp over a large hole in the incoming pipe (which allows water to flow into the tank).
2. Tighten bolts on the upper heater element mounting plate.
3. Install a toggle plug through a small, circular hole in the side wall.
4. Use chains and load binders to latch a gasket and metal backing plate over a large, irregularly-shaped hole in the side wall.
5. Install a 1″ pipe repair clamp over a hole in the incoming pipe.
6. Install a full-opened gate valve onto a pipe nipple on top of the tank, then fully close the valve.

Before using the leak monster in training, load the tallest pipe with at least six brightly colored table tennis balls (Figure 8.6). If your trainees perform the required steps correctly, they will be rewarded by the sight of a surge of water blowing the balls out the top of the pipe as the final repair is completed.

Figure 8.5   The leak monster allows practice in containing several types of liquid releases.

V. 150-Pound chlorine cylinder

An actual 150-pound chlorine cylinder serves as the basis for this device. The cylinder we use was donated by a local supplier after being taken out of service. Pressurizing a container of this type allows trainees to practice various chlorine repair operations. The technique described here could also be used to pressurize a ton container.

The cylinder is pressurized with air through a fitting installed in the side wall (Figure 8.7). To pressurize your cylinder, begin by drilling a hole in the side wall and tapping it to accept a $1/4''$ NPT (national pipe thread) fitting.

For obvious reasons, insure that any container previously used to store hazardous materials is empty before drilling, cutting, welding, or doing anything else to the container.

Note the fitting we used for attaching the airline to the cylinder (Figure 8.8). This is a universal swivel fitting, so that the hose always points downward for minimal interference.

For pressure, use whatever source of compressed air is available. We use SCBA cylinders that supply air at approximately 2200 psi. We use a two-stage regulator to reduce the cylinder pressure to 30 psi (Figure 8.9). A "T" in the airline allows us to pressurize both the 150-pound cylinder and the

LEAK MONSTER SCHEMATIC

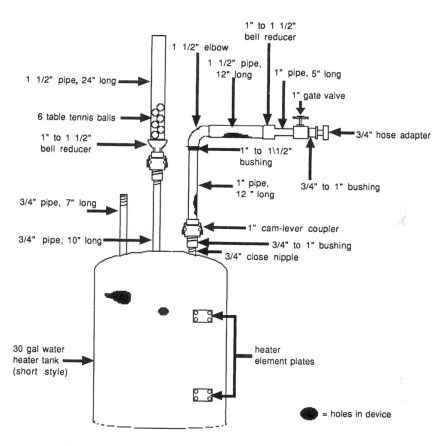

Figure 8.6 The general containment device was built using a junk water heater tank and parts from the local hardware store. (Courtesy of D. Alan Veasey, from *Fire Engineering*, 149(11), 69, 1996.)

ton container from the same source. All connections are fitted with quick connect couplers for ease of assembly and disassembly.

In order to protect your trainees, always use a regulator to reduce the pressure from high pressure air supplies such as air cylinders. Even at reduced pressure, maintain an instructor presence at all times during training to prevent trainees from committing unsafe acts.

Before attempting repairs, trainees need to learn how a chlorine cylinder is constructed. They must become familiar with the tools and equipment provided in the repair kit.

Instruct students to perform repairs in teams of at least two, since that is the way they will work in an actual emergency. Before the arrival of each team, create several leaks for repair. Simulate leaks by loosening fittings

Figure 8.7 An actual 150-pound chlorine cylinder serves as the basis for this device. Compressed air is used to simulate escaping chlorine gas.

(such as the fusible plug, packing nut, outlet cap) and opening the valve. Likewise, loosen the entire valve assembly slightly from the cylinder.

Provide the students a spray bottle of soap solution to use for leak detection. The bubbling of the soapy film over points of leakage substitutes for the use of an ammonia solution for leak detection in an actual chlorine repair operation.

Since trainees will be repairing a 150-pound chlorine cylinder, provide them a Chlorine Institute Series "A" emergency kit to use in repair operations (Figure 8.10). Initially, let them do simple repairs, such as stopping leaks in the valve region by tightening fittings or closing the valve.

Next, remove the outlet cap, open the valve, and instruct them to proceed as if they have a leak that the previous methods did not stop. They will be required to install the hood assembly provided with the "A" kit over the cylinder valve. As another option, equip the cylinder with a valve that you have intentionally damaged as a way of requiring installation of the hood assembly.

Figure 8.8 The universal swivel fitting allows attachment of an airline with minimal interference to the mobility of the cylinder.

VI. Chlorine ton container

The basis for this device is a ton chlorine container head that is mounted on a steel frame and fitted with casters, carrying handles, and a hoisting ring (Figure 8.11). The training device can be purchased from a commercial supplier of chlorine training aids. Such a device is also available with a waterline connection to the liquid valve and lower fusible plugs.

Pressurize the vapor valve on the ton container training aid by installing an airline fitting in the upper eduction pipe (Figure 8.12). Pressure for this device is supplied from the same air supply and regulator used for the 150-pound cylinder.

To use the ton container head in training, follow the same general procedure described for the 150-pound cylinder. Major exceptions are that the fusible plug arrangement is different, and the Chlorine Institute "B" kit is required for ton container repairs.

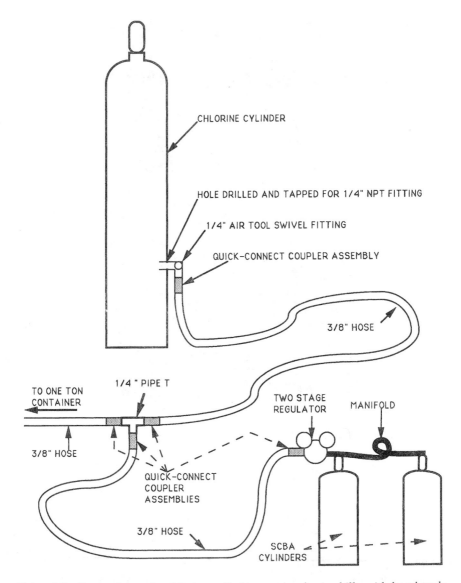

Figure 8.9 Pressurizing the chlorine cylinder requires basic skills with hand tools. The regulator is required to reduce the pressure to a safe level for trainees. (Courtesy of D. Alan Veasey, from *Fire Engineering*, 149(11), 70, 1996.)

Remember the considerations related to pressure hazards discussed for the 150-pound cylinder. These will also apply when training with the ton container head. Therefore, observe the same safety precautions.

Figure 8.10 The pressurized 150-pound cylinder provides realistic practice in containing releases from gas cylinders.

VII. Considerations for implementation

When you train with these or other devices, as you do at all times during hands-on training, give the safety of your trainees the utmost consideration. For example, when conducting training with pressurized containers, regulate the pressure to a relatively low level. Caution your trainees about pressure hazards and observe them carefully to be sure they do not take actions that might put them at risk. Also, consider area safety hazards, such as muddy footing conditions, that can occur when training with devices that use water.

Figure 8.11 A commercially available training aid created from a ton container head serves as the basis for this device. It has been modified so that it can be pressurized with air.

The devices described here are versatile. They can be used in simple exercises intended to teach basic containment skills, or incorporated into complex exercises such as mock incidents. Give full consideration to the training objectives anytime devices such as these are implemented into your training program. For example, none of these devices is appropriate for training below the technician level since awareness- and operations-trained responders should not approach the point of release.

Prerequisites for using hands-on devices are an important consideration. For example, assume that you intend for technician trainees to repair leaks on various devices in a field exercise while wearing a full ensemble of protective equipment. In that case, be sure that requisite objectives, such as containment procedures and use of PPE, are achieved earlier in the course. This should keep trainees from being overwhelmed by too much new experience at one time.

The designs of the devices used as examples in this chapter were adapted to the equipment and supplies available to us. Think of the plans included here mainly as examples. Feel free to modify or replace them as needed based on the needs of your trainees and the tools, equipment, and supplies available to you. Remember that your ultimate goal is to create devices that will allow your trainees to practice the types of procedures they may be called on to perform in the real world.

CHLORINE TON CONTAINER TRAINING DEVICE

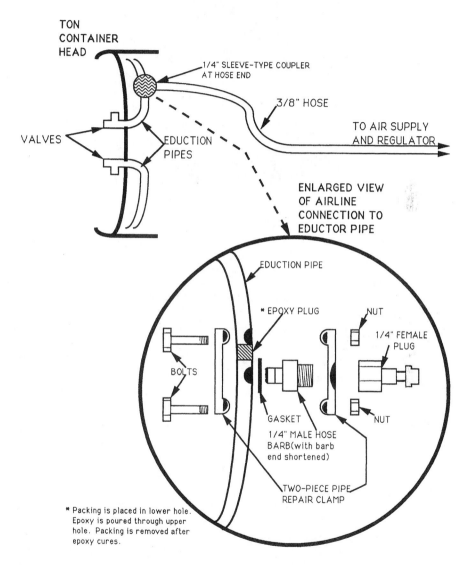

Figure 8.12 The ton container head was modified by installing an airline fitting in the upper eduction pipe, allowing compressed air to simulate escaping chlorine gas in training. (Courtesy of D. Alan Veasey, from *Fire Engineering*, 149(11), 71, 1996.)

Interview

I'm Judi, a labor educator. I started teaching safety and health with a union committee that meets once a month at our university labor center. That led to training for other unions. I teach safety and health, in part, to give people the tools to argue grievances, investigate accidents, and do their own research — and also as a way to organize their members.

Safety issues are probably among the most prominent in internal organizing drives because they are so visible. The people in the class are not the ones who can make the changes — all they can do is try to bargain for changes. That produces a lot of frustration. There is also frustration with their members, who are like all of us: as long as things are going along okay, we don't worry about safety, and we don't want to talk about it. I build some time into every class to let students vent their frustrations with their members, their leadership, the world, and management — certainly management most of all. They get it out of their systems and we go on.

I teach two sets of skills that are common to all labor education. One is negotiation: regardless of what the issue is, whether you are on the safety committee or the bargaining committee, you need to be able to negotiate. The other is organizing: you have to be able to get your membership behind you. What I add is a knowledge base of safety, so the students know something about issues, about OSHA, and about other regulations to help them negotiate and organize. Safety issues are important because you're talking about lives. It's not the same as a discipline case, where you're giving them the equipment to argue *just cause*. It's important to save that one individual worker's job, but with safety you're saving a limb, or a life, or saving a whole lot of people from getting hurt or sick.

Good adult educators know they are going to learn more from their class than the students learn from them. None of us purports to be an "expert." We have technical knowledge, and we have certain legal knowledge, but at heart we recognize that our students know more about where they are and what their experiences mean than we do. There's a respect there. We know they're just as much "experts" as we are — we just need to give them some information they can use to do their jobs better. We have to be very careful what we say, and very respectful, and very cautions that we don't say, "Oh yes, absolutely, go do that," because we don't have to live with the consequences. We leave the plant or the mill and go back to the university at the end of the day.

I enjoy teaching. It's fun for me. Adults are fun to teach. If you have good sense of humor with them, and roll with them, they're just plain fun.

chapter nine

Using computers in training

The computer has come to stay. The positive impact of computers on our daily lives is enormous. Computers even improved the way we wrote this book, with word processors and computer illustrations. In this chapter, we will look at a few of the ways that computers can enhance and improve training, as well as a few cautions about potential pitfalls along the way.

We should point out a bias on our part. We enjoy and encourage the use of computers as tools in nearly every aspect of developing and doing training, but a computer will always be just a tool, not a replacement for the trainer. We hope you have seen in this book how a prepared and enthusiastic trainer can draw the participants into a class and keep them involved. All the glitz and "gee whiz" stuff a computer can bring into a class can't replace the human interaction between participants and trainers. Remember, the trainer's objective is to communicate information effectively, not just to dazzle the trainees.

With that said, let's look at some of the wonderful ways a computer can enhance the training experience for both the trainer and the trainees. We'll look at using computers before training, in the classroom and as the classroom. First, let's talk about some important basic computer terms and concepts.

I. Computer basics

Computer technology changes faster than any other technology humans have invented. Early computers were huge, simple machines that filled entire rooms and buildings. Today's computers are faster, smarter, and can fit in the palm of your hand. Any discussions of computer specifications and details would likely be outdated by the time the book is published; however, a few concepts are important.

A. Personal computers and networks

The computer you have available may be a stand-alone personal computer or a workstation on a network. The personal computer is likely either an

Apple (Apple Computer, Inc., Cupertino, CA) or an IBM compatible (International Business Machines Corp., White Plains, NY) often called a "PC." These two "breeds" of computers account for almost all of the personal computers in use today. Programs are written for either type, so you should be sure that you buy the proper version of software for the computer you use.

The personal computers store and process information independently of other computers. The primary components of a computer are:

- The central processing unit (CPU)
- Random Access Memory (RAM)
- Storage devices such as floppy disks and hard drives
- Input devices such as keyboard and mouse
- A display monitor

Multimedia computers also have a sound card, for processing sound files, and a compact disc — read-only memory (CD-ROM) drive. Early CD-ROMs could not be modified by the computer user, hence the term "read-only memory." Lately, recordable CD-ROM drives are available that allow users to create their own CDs that can store tremendous amounts of information. Other accessories and devices may be added to the computer for specific functions.

A convenient type of personal computer is the laptop or notebook computer, which can be taken easily into the classroom. Laptop computers have all of the components of desktop personal computers, but they are packed into a small, transportable case. Because they can be so easily carried from place to place, laptops are ideal tools for a trainer to take into a classroom (Figure 9.1). The laptops have ports that allow trainers to connect video display projectors and other accessories.

Multiple computers used in the same building or facility may be connected together by a network. The purpose of networking computers is to allow them to share information and resources. In many cases, the network consists of personal computers that are simply linked together. In other cases, a server or central computer stores all programs and files and accesses these files from simple workstations called clients. Computers set up for a network may not be transportable to the classroom, but are useful for materials preparation, as we'll discuss later.

B. The internet

A truly amazing development in the world of computers is the Internet. The Internet allows computer users to connect to other computers around the world. It also allows access to information any time of the day or night. We will be talking about some of the resources available over the Internet later in this chapter. To take advantage of these resources, the trainer must be connected to the Internet. If your computer is on a network, you may have

Figure 9.1   Laptop computers move easily into the classroom, and have many uses.

a direct connection to the Internet. Your network administrator can tell you if you are connected to the Internet or not.

Computers that are not on a network can still access the Internet by means of an Internet Service Provider (ISP). The ISP provides a local telephone number that your computer can dial, using its modem to connect to the Internet. The ISP charges a monthly fee and may provide some other services, like giving you an e-mail address and a free World Wide Web page.

The World Wide Web is the most popular part of the Internet. A web page is a collection of text and graphics laid out as a whole page. A web site is a collection of pages produced by the same company or organization that are linked together. A user can access the Web with special software called a browser. With the browser, the user simply points the mouse to a picture or text, known as a hyperlink, and clicks the mouse button to move from one page to another. The links can point to other pages on the Web site or to pages on other Web sites around the world.

II. Using computers to prepare for training

Trainers today have access to a mind-boggling amount and variety of electronic resources for preparing to deliver training. They can research topics quickly and conveniently. A computer and color printer can generate impressive and effective visual aids. They can communicate and collaborate with colleagues around the world. All of this is possible without even leaving their office.

A. Research and reference

One direct result of the computer revolution is that information can be stored, retrieved, and distributed electronically. Where once we had to possess or borrow a printed book, journal, or article to learn about a topic, now we can literally view whole libraries through our computer screen. The information contained in tens of thousands of material safety data sheets can be stored on a single CD-ROM (Figure 9.2). Information about safety products can be acquired from vendor websites at any time of day or night. You can search entire databases for information about any topic.

Figure 9.2   A single CD-ROM holds a huge amount of data.

What information would help you understand a topic more fully? Chances are there are many sources available to you through your computer. For example, you must read the OSHA Hazard Communication Standard (HazCom) to prepare training on that topic. There are CD-ROMs that contain the entire set of OSHA regulations. You can also go to OSHA's website and view the regulations for free. Do you want to know how OSHA interprets the HazCom labeling requirements as they pertain to shipping containers? These same computer resources also contain all of the letters of interpretation OSHA has sent, as well as compliance directives and other reference material from OSHA.

You can learn the properties and hazards of methyl ethyl ketone or any other chemical from material safety data sheets that are available from commercial software companies or at no charge from some Internet sites. There are companies that provide information on CD-ROM or over the Internet for customers who buy a subscription. Other companies sell large databases of health and safety information on CD-ROMs that are easily searched through by the computer.

B. Developing classroom materials

Trainers can use the computer to create eye-catching overhead transparencies that will reinforce their message. (See Chapter 5 on training methods.) You can use presentation software to type your information into preformatted templates. You can also add pictures or graphics called clipart, available from commercial software packages. After you have developed the presentation, you can print transparencies on a color printer. Or, as we'll discuss later in the chapter, you can give your presentation directly from the computer.

You can also create attractive handouts, fact sheets, information packets, or other printed material with easy-to-use word processors or more sophisticated desktop publishing software. If you use color in your documents, remember that you'll have to print out enough copies from your printer, or pay a commercial copy center to copy the pages for you.

C. Communication and collaboration

One of the most popular uses of the Internet is sending and receiving e-mail. E-mail can be sent from person to person or distributed to a group. Two or more trainers in different geographic areas can collaborate on a training project via e-mail. Whole documents can be shared by attaching them to e-mail or by using File Transfer Protocol (FTP) sites. E-mail eliminates the need to try to synchronize schedules or "play phone tag" to share ideas and information

Trainers can also subscribe to e-mail mailing lists or listservs that may contain thousands of participants with similar interests. E-mail messages posing a question or making comments are automatically sent to everyone on the list. Each person on the list may then read the question and post a response to the whole list or may e-mail the response directly to the original author. For trainers, this means access to thousands of other training professionals who can share their talents and expertise.

Many companies, professional societies, and government agencies have e-mail addresses to which anyone can send a request for information or technical assistance. Because e-mail can easily be forwarded, a question is more likely to make its way to someone who can provide an answer, even if they are in a different organization than the original recipient.

III. Using computers in the classroom

Take the computer into the classroom and you'll be amazed at the different uses you find for it. It can be used for delivering a presentation or for reference during small group exercises. It can be used for interactive methods, such as games and scenarios. Today's computers can store so much information that there really is no limit to the different kinds of programs and resources you can bring to the class.

A word of caution is appropriate. You don't have to work with computers long before you experience a problem that stops your computer cold. In your office, this is often just an inconvenience. However, if you are depending on the computer as a major part of the training course, a computer problem becomes a showstopper. There are several things you can do to reduce the likelihood of such an occurrence.

- Use a better quality laptop computer for in-class work. Price is not necessarily the best indicator of quality, but careful shopping and research can identify computers with a reputation for quality and dependability.
- Keep unnecessary software and programs off of the classroom computer. Some programs install files that can affect other programs. Limiting the programs that are installed onto the computer will reduce the likelihood of these conflicts.
- Consider using utility programs on the computer to catch problems and fix them. There are many of these programs available, but some are better than others. Again, research carefully to find the program best suited to your needs.
- Set up the computer and video equipment as early as possible to give yourself time to fix problems or come up with alternatives. Try the computer and projection equipment together before setting them up in the classroom. You may have to experiment with settings on either device to get the proper display and you don't want to use precious class time to do this.

A. Presentation

The software used to create overhead transparencies is also able to create and control onscreen slideshows or presentations. These presentations can vary from simple electronic "slides" to full multimedia productions. The slideshow typically displays information arranged into frames like 35mm slides, except they are projected onto a screen from the computer by way of a special video projection device (Figure 9.3). The computer version allows the trainer to add special effects such as bullet items in a list or graphics that appear suddenly or travel across the screen. Sound can also be added to the presentation to make a point.

Such a presentation, with sound and motion, is often referred to as a multimedia presentation. The term multimedia originally referred to presentations that incorporated slides, videotape, sound, and information from other media into a single orchestrated event. Now it is used more broadly to describe any presentation that incorporates sound, text, color, and motion, even if it all originates from a computer. Multimedia presentations involve more of the trainee's senses, and can stimulate more interest and hold his attention longer.

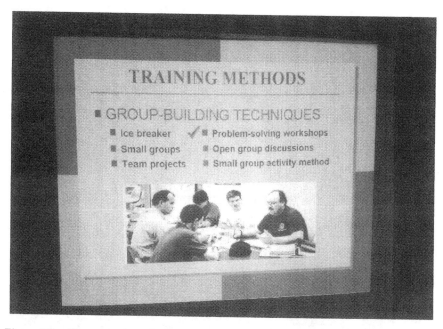

Figure 9.3 Computer-based presentations can be projected onto a large screen, and are convenient to use.

Audiovisual pizzazz is no substitute for good content in training material. Accuracy and relevance to the training group are still vital to effective training. But you can use multimedia tools to communicate your message more effectively, especially to a generation of trainees that grew up watching television and playing video games. Just be careful to keep the pace of the presentation appropriate to the group. Too much onscreen activity may actually be distracting to some and may cause the message to get lost in the "show."

B. Resource for class exercises

A trainer can use computer resources in small group exercises, just as she would use printed materials (Figure 9.4). The resources that are useful in preparing for training would also be helpful to trainees in problem-solving exercises. Since computers are becoming more commonly found throughout most workplaces, it is appropriate to train all workers in how to use them to find information. The most likely limitation to this use of computers is the budget to purchase the computers and software for use by several groups at the same time.

Examples of software that can be used in these exercises include:

- MSDS databases stored on CD-ROM
- Regulation databases that provide access to the regulations

Figure 9.4   Computers have a variety of uses in small group exercises.

- Safety plan-writing software that produces a safety plan based on information about the site conditions and hazards developed by the trainees
- Emergency response management software such as the National Safety Council's Computer-Aided Management of Emergency Operations (CAMEO®, National Safety Council, Itasca, IL) or their companion software for mapping contamination plumes called Aerial Location of Hazardous Atmospheres (ALOHA®)
- Other health and safety information databases.

Small groups may access information resources on the Internet like the ones discussed in the section on materials development. Depending on the training facilities, access to the Internet will be through the company's network connection or through a local ISP. Once access is set up, small groups can gather information from resources around the world. There can be problems with getting and keeping an Internet connection during the class, such as busy signals at the ISP's access phone number. Trainers should shop around for a reliable ISP and be sure to give plenty of time to check the access before class starts. This is especially important when training at a hotel or rented facility because the phone lines may need to be switched on or changed to allow outside calls.

   Trainers who use computers as classroom resources must be sure that at least one person in each group is familiar with the basics of operating a computer. There may need to be a brief orientation to the whole class if it seems that many of the trainees are not familiar with computers (Figure 9.5). Also, it will be important to have extra instructors available to troubleshoot computer problems and keep the groups from getting stuck.

Figure 9.5   An orientation session may be necessary before asking trainees to use the computer.

   Trainers may have difficulty keeping the class from exploring other programs on the computer, thereby being distracted. This is another reason to heed our previous warning to limit the programs that are installed on the classroom computers. Another option is to use passwords to control access to various programs that are installed. You'll want to remove any games that may have come as part of the operating system software for the computer. The presence of roving instructors can also keep groups on track.

   Controlling access to Internet resources poses a special challenge. Because the World Wide Web is so easy to use and navigate, trainees may find it particularly tempting to wander off track. If you are concerned about trainees accessing objectionable material while connected, you may want to invest in filtering software that recognizes and blocks access to objectionable sites. The programs use keyword searches to look for words on the site that warn of adult content, hate speech, or other objectionable material. These programs are not foolproof and may block desired sites that contain certain keywords, but most can be customized by the user and can be set to allow access under certain conditions.

C. Software demonstration

Some trainers use computers in the class to demonstrate examples of different kinds of software. This is especially useful in update or refresher courses as a way to keep trainees informed about resources that they might want to acquire for their workplaces. The software company usually recognizes such a use as a marketing opportunity, and the trainees get the opportunity to try out products in a hands-on, low-pressure setting.

It's important to remember that software companies are in the business to sell programs, not give them away. If you are approaching a company about giving a demonstration, be prepared to show how it could generate sales of their product. They often will provide a free demonstration copy or limited version to the trainer or may provide a full version for a limited time with the understanding that you will make product literature available or provide the company with contact information about the trainees. Limited versions usually run like the full version, except key functions such as printing or saving are disabled, or the database may have only a few chemicals instead of the full database.

Many of the "demo" disks offered by software companies are simply computerized slideshows or tutorials that tell about the product, but don't let users really work with the program. Be sure that what they are sending you is a functional demo that will let users actually try the various features of the software. The hands-on use of the software lets the trainees apply the software to their needs and will keep their interest in the classroom.

D. Computer-based games and simulations

Trainers have long found it helpful to bring games into training in certain circumstances or settings. Now, there are companies that sell computer programs that allow trainers to create customized games based on television game shows or popular board games. These games take advantage of trainee familiarity with the premise of a game and allow the trainer to write questions that are specific to the topic of interest. One software company sells a multi-game product that includes a Jeopardy-like game. It includes easy-to-operate gameware, colorful graphics, and a choice of regular or wacky sounds to denote right and wrong answers. Questions and answers are added by the trainer, and a number of different games can be saved to fit different courses. Trainees love playing these games (Figure 9.6).

Figure 9.6 Computer-based quiz games include clever sound and bright graphics; this team begged for more.

Computers can also be used to create simulations, often using the concept of virtual reality. Virtual reality is computer technology that allows the creation three-dimensional areas and permits the users to move through the area, and manipulate and see various sides of objects. This sophisticated technology requires a significant investment of resources and expertise. One training organization has created a simulation of a drum storage room that they use for hazard assessment exercise. The trainee can "walk" through the room on the computer and identify different hazards.

IV. Computers as the classroom

Computers also serve as the classroom itself. The two most common examples of this are interactive computer-based training (CBT) and distance learning. Advocates for the use of each of these approaches to training point to the efficiency of letting trainees come to training on their own schedules and at their own pace without being tied to a classroom. As you might expect, there is controversy associated with the effectiveness and appropriateness of the sole use of computers for training. Since this book is focusing on preparing trainers, we'll keep our discussion of these approaches brief.

A. Computer-based training

Computer-based training uses the computer to deliver information to an individual trainee in a self-paced format. The information comes in the form of text, video, sound, graphic or other multimedia resources. The trainee views the presentation and then responds to questions that test her retention and understanding of the information. The training is often broken into short sessions that may be stopped and started at any time. The trainee can progress to the next topic at any time or may be forced to complete the topics in a sequence, depending on how the program is written. Training is complete when the trainee correctly answers a predetermined number of test questions. When the trainee gives the wrong answer to a test question, she is given a chance to review the information before testing again. The computer is able to keep a record of the entire training process and print out reports of the results for individuals and groups of trainees.

Some advantages are immediately apparent. Because trainees can start and stop at any time, they can work the training into a busy production schedule. A computer training center can be set up near the shop floor in a factory and trainees can be given time for training without having to take a whole day off to sit in the classroom (Figure 9.7). Also, the trainee can take a break when he feels that he is losing concentration. If he is having difficulty with a particular topic, he can repeat the lesson as often as he wishes.

Advocates of CBT point to the variation of training skills between trainers and the fact that even professional trainers have bad days as potential consistency problems with traditional classroom training. The computer program ensures that all trainees receive exactly the same training. Experienced trainers view the "one-program-fits-all" advantage with skepticism. But the consistent information, coupled with a human trainer to adapt it to specific trainee needs, may be advantageous.

We also see some disadvantages to the sole use of CBT. Computers are simply not able to respond to questions effectively. Human trainers can hear a question and interpret what the trainee is asking. Computer programs, on the other hand, require the trainee to know how to look for an answer among a list of topics. Even if the topic for the question is found, the wording of the information may not fully answer the question. Human trainers can discuss the question with the trainee until he understands.

CBT is considered interactive — to a point. The trainee is able to navigate through the program and choose the training topics of interest, thereby controlling his training experience. But his body remains stationary throughout the training. Also, he doesn't have the opportunity to interact with his peers and the trainer and learn from their experiences. This may result in less practical application of the material that is learned. In the words of one trainer who advocates using live instructors and actual hands-on practice, "You can't teach a worker to drive a forklift without putting him on a forklift."

Figure 9.7 Computer rooms can be set up close to the shop floor for the convenience of workers.

The Occupational Safety and Health Administration (OSHA) has been asked to provide some guidelines for CBT use. At the time of the writing of this chapter, OSHA's interpretation of the refresher training requirements of the Hazardous Waste Operations and Emergency Response (Hazwoper) standard stated that CBT alone would not be sufficient. OSHA recognizes CBT as a valuable tool, but is concerned about the issues of answering trainee questions and the evaluation of hands-on skills.

B. Distance learning

The correspondence course has come to the computer age, and is now referred to as distance learning. The development of e-mail, the Web, and other Internet technologies makes it possible to take courses and even pursue a degree without setting foot on a college campus. This high-tech educational format takes different forms. One form is "asynchronous" courses in which messages from the instructor and classmates are uploaded (sent to) and downloaded (received) from a central web site or server as simple text files that can be read and responded to. Students can access a "class" at any time whether or not other students are on-line at the same time, thus the term "asynchronous." Another form is the "chat-room" format in which class participants are all logged onto a Web site that allows each participant to post text messages that are then viewable by everyone in a scrolling screen. The instructor may pose a question or make a statement and then all participants can respond with comments. The truly high-tech version of distance learning called video conferencing uses live video and audio transferred over the Internet.

Distance learning involves interaction between a trainer or facilitator and the trainee, overcoming one of the disadvantages of CBT. However, for asynchronous courses the interaction is delayed because the trainer must check for messages and then send a response to the trainee who then must check for messages to receive it. The format requires communication skills on the part of the trainee to form the question in a clear way, or repeated clarifications may be needed. This format is used primarily for online academic courses, but not frequently for training unless coupled with a CBT course.

"Chat room"-type courses eliminate the delay in answering questions. However, they are limited to primarily text on a screen, not the most effective training method. The format can be used as a means of communicating limited amounts of new information, but should not be used as a primary training method.

Video conferencing offers the next best thing to actually being in the classroom. The immediate interaction allows discussions between trainers and trainees. Trainers can offer guidance from a remote location without having to travel. However, the equipment and technology involved is substantial and is usually too expensive to be used for training in all but the largest of companies or facilities. It does not allow for effective hands-on training with equipment, so it would not be appropriate for all types of courses, but for some situations is an alternative that can be considered.

V. Summary

In this chapter, we introduced some ways that trainers can use computers and the Internet both inside and outside of the classroom. Even as you read this chapter, you are undoubtedly able to identify other uses of computer resources in and for the classroom. There have probably been significant changes in technologies and new tools are available. We encourage you to keep your mind open to other ways you can use new technologies to improve your training. But be careful, because "all that glitters is not gold." Be sure that any tools you use in training help you achieve the objectives of the training or you may simply be creating an entertaining waste of time.

Interview

I'm Johnny, and I train for the National Safety Council. I think a couple of things are critical in training. The first is knowing your audience, and modifying your presentation based on them — their education, interests, whatever. I have people introduce themselves, and I ask everyone, "Why are you here?" This gives me background on them, and I've gotten them to participate so they'll be more likely to open up later in the class.

I ask classes what single word they think most people use when describing safety meetings or training, and the word that always comes up is BORING. They expect to be bored. Use something to get their interest. I love to use humor — cartoons, gimmicks. I want them to wonder, "What's he going to do next?"

I have heard it said that people remember 10% of what they read, 20% of what they hear, 30% of what they see, 50% of what they see and hear, and up to 80% of what they participate in. So audiovisuals are important, and so is getting people to participate. That's how people learn best.

Some of the best teachers are preachers because they've mastered presentation skills. They are great presenters. The volume of the voice changes — you've been in churches where the pastor is pounding on the podium to make a point, and the next minute he's leaning forward and speaking in a whisper, and you're leaning forward to hear it. They are masters of going from humorous to serious. Those are skills a good teacher, a good presenter, should utilize. Many people have a lot more knowledge than I do, but just don't develop the presentation skills.

I tell people I'm not very smart, but I learn from other people. I find out who knows all about what I need to know, and I go and meet with them and pick their brains. I talk a lot, but I also listen. Everybody I know knows a lot more about something than I do. I don't mean to be cocky or arrogant, but I believe if you said, "Johnny, I want you to go the AMA convention next year and give a one hour presentation on the latest techniques in brain surgery," I could do it. And do a good job, because what I would do is seek the most knowledgeable people and pick their brains. I don't have any knowledge of brain surgery, but I guarantee I could find the right people and quote them, and make the presentation interesting.

Basically, I'm a ham. I enjoy training. I study what I have to present, and I really prepare for it. The new fork lift standard, for example. I studied it for weeks, met with OSHA to get their input on the training material I developed, and asked the local OSHA director to come to the first class and critique me. I get really nervous when I do something the first time, but part of getting over it is preparing well.

chapter ten

Evaluating training

You have designed, developed, and delivered training. You worked long, hard hours to get it all accomplished, and everybody involved agreed it was a great success. Congratulations! But wait a minute — what is "success?" To know whether you achieved it, you must define it. Once defined, successful training can be measured in a variety of ways.

I. Defining success

We will let you write your own definition, and we will show how other people define successful training. You will notice that their definition focuses not on the training, but on the trainee, as you do when you write learning objectives.

Most trainers define the success of training according to answers to one or more of the following questions.

- How did trainees respond to the program? Did they like the material and the trainer? Did they feel the instruction was a valuable way to spend their time? Did they believe it was related to their jobs?
- Did the trainees learn what they were supposed to learn? In other words, did each of them achieve the learning objectives?
- Have they transferred what they've learned to the job? If not, be sure your learning objectives fit what they need to be able to do on the job. If accomplishment of the objectives does not lead to safer on-the-job actions, then perhaps you should revise the objectives and training program content.
- Have skill levels been maintained over a period of time? Perhaps you should retest at regular intervals after training.
- Does the training bring the company into compliance with the regulatory agency or accreditation organization that furnished the motivation and guidance for the course? Regulatory compliance is the most common trigger for training.

- Do the results of training justify the cost? Most companies want some indication that training is having a positive impact on the organization as well as on the individual trainees.

II. Evaluating success

The process of determining whether your training is successful is called evaluation. Without evaluation, your training program is incomplete. This chapter is designed to help you answer five questions: Why, what, who, and when do evaluate, and how to use the information once you have it.

A. Why evaluate?

There are several reasons evaluation is an important part of your training program.

1. Measure success in attaining objectives

You want to know how you are doing, whether trainees feel their time is well spent, and whether they are learning what you set out to teach. In other words, whether the learning objectives are being met. If they are not, and you are sure you wrote them accurately, then either you're content or your methods need revision.

2. Get feedback for improvement

You need feedback to make the training better each time you do it until it is so good it just can't get any better. There is usually room for improvement, and there are several different groups of people who can help you improve.

3. Establish cost effectiveness

Whoever is paying the bills deserves to see evidence that the program is well run, the training is effective, and you are responsive to your evaluators.

4. Improve your job prospects

You gain employment value as a good trainer. Keep summaries of your best evaluations; you never know when they might be helpful. Use them when you seek a better job, or more responsibility.

5. Recruit for courses

If any of the training you do is voluntary on the part of attendees, or if you are a trainer for hire, publish comments from people previously trained. With their permission, you can use photos, names, and company affiliations. Without it, you can reprint their comments anonymously.

B. What will you evaluate?

When we think about evaluations, the first thing that comes to mind is probably the form trainees fill out just before they leave at the end of class. The form is important, but certainly does not constitute a complete evaluation of the course.

Your evaluation should focus on four areas. Each of these will require a different measurement. Educators call tests that measure items such as these "instruments." Some instruments measure the process of training, that is, what happens in the class, and some evaluate post-training outcomes.

1. Process measures: trainees' opinions

Some instruments evaluate things that happen during training: trainees' reactions to the program in general including the trainer, the training environment, and the skills being taught.

a. Instrument. A student evaluation form is used for this. Questions must be worded carefully and tried out to see if the group you are training reads them as you intended, or if they thought you meant something else.

b. Guidelines. Be as straightforward as possible. Use short words and very brief sentences or phrases. Make sure the directions are at the top, the actions you want trainees to perform are well explained and consistent throughout the form, and the font is large enough to read. There should be plenty of space for comments if you ask for them. Make clear what comments you request; for example, if you ask, "Were topics too long or too short?" the answer may be "Some too long, some too short" and you will never know which were which. If you use a numerical rating system, define what the numbers mean.

Don't ask workers to evaluate items they are not prepared or qualified to judge. For example, "Was the trainer well prepared?" may not be obvious in class. Do you mean to ask if she was well organized, or whether she knew the material, or if she ran out of handouts? Asking "Was the information accurate?" is foolish if the reason you are training is to provide information not already known by the students. How do they know if it is accurate or not?

A good evaluation measures more than the personality of the trainer. The authors have watched poorly informed but energetic trainers get good student evaluations while teaching dangerously incorrect facts, while accurate but less charismatic trainers scored poorly. Evaluations should detect two important factors, accuracy and interesting delivery, but don't ask students to rate instructors out of context of the material they teach.

There is a place for instruments that measure only a trainer's performance. Good training organizations constantly seek to improve their trainer's skills, and may do annual program quality evaluations. Figure 10.1 shows one such instrument. Evaluators usually are lead instructors or program manager, not class members.

INSTRUCTOR EVALUATION for PROGRAM QUALITY CONTROL

Instructor_____ Date_____

Course_____ Topic_____

Was the lesson plan fairly closely followed, including estimated time?

Did instructor appear to have sufficient knowledge of the topic?

Did instructor make the purpose of the lesson clear, and relate it to the course goals?

Was instructor's manner friendly, interested and interesting, and did it set a good climate for learning?

Was instructor's delivery of information clear?

Were students actively participating?

Were questions answered, or referred to others, or remanded for future answers when more information was found?

If there was an exercise, were students grouped without confusion?

   Were clear instructions given?

   Was the exercise summarized and related to course objectives?

(Use the back for additional comments and suggestions for improvement)

Figure 10.1 An evaluation form used by the training institution to help instructors improve performance. This form is not given to trainees.

2. Process measures: trainees' knowledge

It is important to determine trainees' knowledge of facts, techniques, operations, MSDS, etc. This is equivalent to finding out if the learning objectives have been met. For regulatory compliance, it is important to document knowledge.

a. Instrument. Use tests, written or oral, or checklists the trainer fills out as trainees perform certain skills and operations.

b. Guidelines. Tests are strongly discouraged by some organizations and companies, and always used by others. Workers without much education, and many others as well, do not like written tests. They violate several of the principles of good adult education. They make workers uneasy — even fearful — and they make no allowance for poor reading and writing skills. Avoid them if you can.

If you decide a written test is necessary or desirable, first decide what you want to document. Do you want trainees to remember all the facts you presented, or simply to know where to get information they need? If the latter is your goal, ask questions that can be answered using the training materials and are designed as part of an "open book" exam.

If testing job skills is the best way to measure the attainment of the learning objectives, devise a checklist to be filled out by the trainer, or by a partner, as workers perform tasks. File this with other documentation for your course.

3. Outcome measures: changes in behavior

Outcome measures, things that happen following training, are more difficult to evaluate than process measures. Changing job behavior as a result of the training is likely to be your goal, and may be the measurement you prefer to attain. Although you probably cannot show a cause-and-effect relationship between training and behavior changes, you can determine if there is a correlation between the two.

a. Instrument. Self-report (asking the trainees about their behavior changes at intervals after the course) is one way to measure behavior change. You can also request reports from trainees' supervisors, or arrange to observe them yourself.

b. Guidelines. The questions on this instrument must be very specific. Write them to clearly describe the differences in behavior you wish to detect, perhaps with frequency qualifiers such as "always" and "most of the time." You can also fill out a prepared checklist while observing the trainees at work.

4. Outcome measures: economic gain

Economic justification, in terms of reduced accident and injury rates, job turnover, or improvements in the quality and quantity of work are the outcome measures that will most clearly impress the company. Being able to show the cost-effectiveness of training is important in promoting support and continuation of the program.

a. Instrument. If you can get access to OSHA 200 logs, other company medical and reporting forms, insurance claims, or quality control records, try to do so. Again, you can't prove training caused the improvement, but decreased accidents and injuries suggest successful training. Indications of increased production or improved product quality allow you to state that safety and productivity are not incompatible.

b. Guidelines. Be honest. Don't try to demonstrate a relationship that doesn't exist, or to claim results that are not supported by the data. If you have no training in statistical analysis, get some help from someone who does.

C. Who should do the evaluation?

Because you evaluate for different factors, you need several different people or groups to serve as evaluators. Consider asking the following people for input.

1. Trainees

Trainees will tell you whether they got the main points, understood the material, were bored, found the information relevant to their jobs, and will be able to do their jobs better or more safely because of the training. They are an all-important group for doing process measurements.

Decide first what you want trainees to tell you; then word the questions to stimulate those responses. Relevance to job tasks and overall satisfaction with the course can be queried directly; knowledge, understanding, and attainment of learning objectives should be questioned indirectly, and possible can best be done with instruments other than student evaluation forms. Don't ask, "Did you get the main point?" but "What was the main point?" or, better still, ask an information question that checks the knowledge.

2. Other trainers

Invite trainers into your classes to give suggestions for ways to increase the learning that takes place there. Ask them for ideas they have used and found to be successful with similar populations. If you are fortunate enough to be part of an organized group of trainers, set up an exchange program for ideas. The authors have benefited greatly from their association with worker trainers in grantee organizations of the National Institute of Environmental Health Sciences; they teach in universities, labor unions, and COSH groups, and have provided excellent help over the 12 years the grant program has been in effect.

Other trainers can help you "see yourself as others see you." A good trainer will advise you on methods, and on personal habits and mannerisms that you may be unaware of. If personal scrutiny embarrasses you, set up a camera in the back of the room and videotape yourself. A personal note: It is our feeling that habits such as putting hands in pockets, turning toward the screen, and other minor infractions that seem to merit major status in some trainer courses are unimportant compared to the overall tone of the class and the clarity of the information delivered.

3. Technical experts

Technical experts should be invited to your training if you are starting a new topic or teaching in an area where you are not completely confident of your expertise. If you are making factual errors, you want to know it so you can set them straight. Other trainers can help here, also, if they are knowledge-able about the material.

4. Advisory committees

Advisory committees can be very helpful if they are made up of people who know the subject, know how to train workers, and are willing to sit in on training and meet with you to give advice. If you are training workers, be sure the advisors are workers or worker trainers, and not teachers at some other level, such as college professors. Committee members can also help trainers improve the effectiveness of training by offering suggestions on gaining support, adapting courses to current needs, and projecting an image of program quality.

5. Trained workers

Workers who have had similar training at other locations can give you new ideas. It is good practice to observe a variety of training sources, if you can be sure of their qualifications and the quality of their programs. Participants learn something new at every class they attend, even when it is a subject about which they have had previous training. If you can manage the time and expense, take a few classes elsewhere and borrow the good ideas you observe.

D. When should you evaluate?

Timing is important, both for in-class evaluations and those that follow later. If you truly want meaningful evaluations, people need the time and moti-vation to fill them out carefully.

1. In class

When evaluations are given out at the last minute, trainees tend to zip through them so they can leave. Often they give them little or no thought. On forms that ask trainees to circle a number, many students make an all-inclusive circle of an entire column. Maybe they really liked (or hated) all the topics that much, or maybe they just wanted to go home.

When evaluating a class of more than one day, some trainers have found it helpful to take time at the end of each day to request that participants fill out the section for that day on a form provided at the start of class. Memories are fresher when evaluations are done immediately.

2. Follow-up surveys

When surveys are done depends on what you want to find out. Allow plenty of time for reminders if the survey is mailed or distributed for return at leisure. It is not unusual to send two or three reminders and still get less than 50% returned.

E. What will you do after evaluating?

As you design evaluations, have a plan in mind for their use. This allows your questions and format to point toward the results your plan requires. If you simply throw away or file away your evaluations, you have wasted your time preparing and performing them.

1. Use them to document competency

Much of your training will be driven by compliance with standards that require training or competency for workers. The evaluations serve as evidence of having met those requirement.

2. Use them to improve the training

Invite several people who were part of the training (other trainers, if there were any) to meet and comment on the evaluations. Plans to change the training should be made by all the trainers involved. If you are the only trainer, get input from several other people who attended, and whose opinion you respect.

3. Use them to boost your ratings

Use your good evaluations when you recruit for further training. Use them to show management the training is successful. Publicize them to encourage others to become involved with the training and help you meet your training goals. For example, you might include in your information flier the statement. "Eighty-eight percent of the last class stated they can perform their job more safely after the class than before they had the training."

III. Designing student evaluations

It is more difficult than you may think to write questions that will result in the specific information you seek. The people who read and answer them often interpret them differently than you did when you wrote them. To test this, try the questions out on several groups who are similar in education, experience, and job assignments to your target training group. After the evaluation forms are put into use in courses, be prepared to alter them if necessary.

A. Clarification of desired outcome

First of all, you must decide exactly what you want to know. One organization that struggled with evaluation questions started by figuring out just what they wanted to know, and listed the following questions as design guides.

- Did trainees get the main points?
- Did they understand the material? If not, why?
- Did the training relate to their jobs?
- Was too much time given to things they already knew?
- Were they bored?
- Did the workshops and hands-on practice help them learn?
- Will they be able to do their jobs more safely?
- Were all trainees, laborers as well as project designers and managers, being served?

B. Writing clear questions

One of the most difficult things any teacher does, in any educational setting, is to write questions that ask what she means them to ask. The only way to determine this is to pilot the questions with several groups of approximately the same educational and experiential background as the target group. Another difficulty lies in requesting specific feedback; in other words, does each answer refer to the course as a whole, a specific topic, the instructor's delivery, or the doughnuts? If a numerical rating scale is provided, is five high or low, and how high? How low? Does four mean usually or medium-high?

Open-ended questions require adequate space for writing answers. It is insulting to ask, "How can we improve this course?" and limit the writing space to one line. Open-ended questions should closely define the information sought. If you ask, "Could this have been taught in a shorter time, or should we have taken more time?" be sure to indicate, in a multi-topic course, that you wish the response to be specific regarding topics. Request further help on how to shorten it, or what to leave out, in order to discriminate between "I really thought it was redundant" and "I just wanted to get out of here."

C. *Electronically scanned forms*

Several companies make forms that can be fed into a machine for grading based on an answer key. The companies design and print special forms based on questions and answers desired by the trainer. These forms have advantages and disadvantages.

1. *Advantages*

- Evaluations are machine scored, which saves time.
- Statistical analysis can be done by the machine, which reports frequencies of answers for each question, questions most frequently missed by high-scoring students (in the case of tests), or instructors who received the highest ratings.

2. *Disadvantages*

- The forms are expensive to design and print, and there is cost involved in scoring them.
- It is necessary to have access to a scanning machine.
- Some trainees who are not used to using this kind of form are put off by it.
- Once forms are designed, there is no changing them. The evaluation cannot be revised without paying for a whole new set of forms.

D. *Evaluations for trainees who read poorly*

A written questionnaire is useless if the trainees cannot read it. Furthermore, they will not tell you they can't read and will attempt to answer, especially if the answers require only that they circle a number or check a box. In that case, the answers become more than useless, they skew the results and confuse the trainer.

If, by the end of a class, you have not figured out the reading level of your trainees, then you have probably not done a good job of eliciting their feedback. In a class that includes people you believe need help with the evaluation, set up an opportunity for these individuals to be asked the questions by someone other than the trainer (whose feelings they may wish to protect) and write down the answers. Do not force non-readers to request this method of evaluating your course; instead, provide it automatically.

E. *Focus groups*

Forming focus groups is a good way to determine if evaluation forms are hitting the desired mark. General guidelines for using focus groups are as follows.

- The group should be representative of the training population.
- Place the group around a table so all participants, including the facilitator, have equal status.
- Focus group members should volunteer, not be drafted. If you bring in lunch, or offer some other inducement, people are more likely to volunteer.
- Don't include supervisors in groups made up of workers they supervise.
- A neutral setting, away from intimidating location, is best.
- Use basic brainstorming techniques to elicit information. In other words, make no judgmental statements about the responses of group members.
- Ask a neutral party to serve as scribe, writing down the proceeds of the meeting, so the group leader is not distracted by this responsibility. The scribe sits away from the group, and does not interact with the group at all.
- Plan questions and desired outcomes in advance, similar to the manner used for writing learning objectives.

IV. Using tests to evaluate learning

At the end of many training sessions, participants are asked (often required) to sign a paper saying they have been trained. The company uses these to document training. A good trainer wants to know more — to have his or her own documentation that the learning objectives were met. One way to do this is to have trainees take a test.

A. Written tests

As we said earlier, written tests violate several of the principles of good adult education. Tests make workers uneasy — even fearful — and they make no allowance for poor reading and writing skills. Avoid them if you can.

One training group wrote a visual "exam" that is used to measure how much workers learned. The questions ask for some action based on drawings. Each page is copied on an overhead transparency, and the trainer shows and reads the questions to the class as the class responds by circling or otherwise marking on their copy. Figure 10.2 shows one of the questions. The test is referred to by the instructors as an exercise, and given at the end of day four of a five-day course. It is scored overnight for documentation, and used as a learning exercise when the trainer goes over each question the next day. Because of the way it is used, questions do not necessarily have right and wrong answers, but can serve as springboards for discussion of health and safety issues.

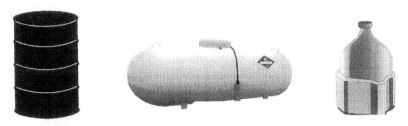

Figure 10.2 Example of a visual exam question. The trainer asks students to circle on their copies the container normally holding pressurized contents.

B. Practical exams

The best way to measure if workers can do the things they have been trained to do is to watch them do the tasks. Use a checklist to document their accomplishments; or, if they have the reading and writing skills to do so, give them a worksheet to fill out as they go from task to task. The practical exam is just like a classroom stations exercise.

V. Follow-up surveys

At six month or yearly intervals after training, some trainers send out surveys to determine how much trainees remember, or to evaluate workplace outcomes of training. Be sure you have permission from the organization(s) to whom you and the trainees report before surveying.

A. Testing memory

Questions for testing memory look much like a written test, and are subject to the same qualifications as tests. The most common forms of questions are multiple choice and yes/no, with a line after each for comments. They can be statistically evaluated by calculating the frequencies of each answer.

B. Evaluating outcomes

Outcomes are the most important goal of safety and health training, and the hardest to measure. They include a variety of changes, including behavior changes, safety improvements, and use of materials provided in class. Surveys ask for self-report answers to questions like "Are you more likely to wear chemical protective gloves since training?" or "Have hazardous chemicals been replaced by less hazardous ones in the process you work with?" Always provide comment lines after each question, since a yes or no answer may not completely explain the respondent's answer. In reporting results, remember that there may be bias in the return of survey questionnaires; trainees who are most enthusiastic and most likely to return them may also be most likely to implement changes.

VI. Examples of evaluation forms

As trainers in an organization with many years of evaluation experience, the authors have failed at writing good evaluations many times. It is with great humility that we present our failures for your examination, along with the reasons these instruments didn't work the way we thought they would. We also present an example that worked, but we didn't design it, as you will see.

Figure 10.3 shows a form once used by the authors. We abandoned the form because the following problems became evident.

- On number one, we were asking if job performance would be better, when what we wanted it to be was safer.
- John Doe always received higher scores than Jim Beam, even though he was a notably inaccurate instructor and frequently told of personal experiences that were untrue.
- Topics taught by Jim Beam were rated much higher than John Doe's. What was being measured? We believed it was animation and friendliness, rather than competency. Charisma is important, but the instrument made it the focus. Effectiveness is our primary training goal.
- The rating scale did not have absolute value. Good compared to what? Who is average?
- Nowhere did we relate topics to course objectives. Topics that included hands-on activities always got better scores, but the others also are required by the OSHA standard. Boring topics such as Review of Standard always received low values, and we did look for ways to make them more interesting.
- Part four was too vague, and really didn't tell us anything.
- In answer to part five, we frequently were told, "Some too long, some too short" without stating which. There was no way to accommodate such a comment.
- The answers to part six were predictable in most cases. Everybody liked the hands-on parts, and we eagerly responded by increasing participatory activities. One told us every year at refresher courses that he liked "the colors" and "the smells" and "the karma." His answers made our day — we loved this guy, whoever he was.

Figure 10.4 represents our attempt to get more useful answers, and to correlate the answers to education and experience. We usually have a wide range of each in this course. Although it was developed by all kinds of so-called experts, the form was graded by focus groups as a disaster and was abandoned. Critics said it was impossible to use, difficult to figure out, had no instructions, and lacked room for comments.

Figure 10.5 is a student-designed form for a training methods course. Beginning and experienced trainers spent several hours considering the same information provided in this chapter, another hour analyzing various

Course Evaluation
HAZARDOUS MATERIALS TECHNICIAN COURSE
Location _____ Date_____

1. After taking this course will you be able to perform your job better?

     Yes   No

2. Please rate the presentations of the instructor

Name:	Very Good	Good	Average	Poor	Very Poor
John Doe	5	4	3	2	1
Jim Beam	5	4	3	2	1
Jane Roe	5	4	3	2	1

3. Please rate the coverage of each topic:

	Very Good	Good	Average	Poor	Very Poor
Hazardous Materials	5	4	3	2	1
Review of the Standard	5	4	3	2	1
Hazard, Risk Assessment	5	4	3	2	1
Site Control	5	4	3	2	1

Etc: All the topics of the course were listed in order.

4. Please the course on the following:

	Very Good	Good	Average	Poor	Very Poor
Interesting, stimulating	5	4	3	2	1
Handouts, material	5	4	3	2	1
Audiovisual material	5	4	3	2	1
Overall quality	5	4	3	2	1

5. Was the time spent on each topic:

Too short?   Adequate?   Too long?
Please comment.

6. Please tell us what you liked about the course.

Figure 10.3 Early student evaluation form, no longer in use.

examples of evaluations, and then were asked to design an evaluation form for process measurement of the course. The form was compiled from the input of six small groups; they were amazingly similar. The class was told before they designed the form that the instructor would use it to improve the course, and they gave her good feedback. The student-designed form would be too long for most classes, where trainees less involved in learning to evaluate would not give it their careful attention.

VII. Further guidance for evaluation

Using the simple guidelines given in this chapter, together with your good sense and the experience you will gain through the process, you can design good evaluation instruments for worker training programs.

The information given here is based on our personal experiences. A tremendous body of knowledge has been compiled on the subject of learning evaluation. If you wish to investigate evaluation theory and practice more deeply, you will have no difficulty finding resources in the education literature.

There are two sources of evaluation guidance specific to worker training programs. One is Appendix E of 29 CFR 1910.120, the Hazardous Waste Operations and Emergency Response Standard. The appendix was written by a group of trainers funded by the National Institute for Environmental Health Sciences; their intention was to structure 29 CFR 1910.121, an accreditation standard for Hazwoper training programs. Instead, OSHA published the document as an appendix to 1910.120. Although the appendix is specific to Hazwoper training, the majority of the recommendations can be adapted to any worker training program.

The American National Standards Institute (ANSI) writes standards for a variety of governmental programs. ANSI has convened a panel of experts to write ANSI Z-490, Best Practices in Safety, Health, and Environmental Training. This standard, which includes recommendations for voluntary application, will most likely be published in late 1999.

Both documents are written by respected, successful worker trainers and can serve as guidelines for the evaluation of program quality. They also make good checklists when purchasing training programs, materials, or contractor services.

24 HOUR CONFINED SPACE ENTRY & RESCUE TRAINING
COURSE EVALUATION

EDUCATION: _____ Grade School _____ High School _____ College _____ Graduate School

WORK EXPERIENCE: _____ Less than 1 year _____ 1 to 5 years _____ More than 5 years

WHERE YOU WORK MOST OF THE TIME: _____ Office _____ Plant _____ Outdoors

	DAY ONE							DAY TWO			DAY THREE		
	Intro to Confined Spaces	Regulatory Requirements	Assessing Confined Space Hazards	Workshop in Hazard Assessment	Selection and Use of Respirators	Respirator Field Exercise	Chemical Protective Clothing	Air Monitoring	Safety Procedures for CS Entry	CS Entry Field Exercise	Overview of CS Rescue	Field Orientation to Rescue Equipment	Field Exercise in CS Rescue
Was this part of the course interesting?													
Did you have the chance to really take part?													
Were you able to follow what was taught?													
Did you learn things that can help you stay safe and healthy on the job?													
If no, why not?													
-Couldn't understand the material													
-Already knew all I needed to know about the topic													
-Information did not pertain to my job													
Should we take more time to cover this material?													
Could we have covered this material in a shorter time?													

Figure 10.4 Revised student evaluation form, no longer in use.

	YES	NO	NOT SURE
Overall, do you think this course was worth your time and trouble?	YES	NO	NOT SURE
Did all topics covered relate to each other?	YES	NO	NOT SURE
Did you feel welcome and comfortable?	YES	NO	NOT SURE
Please tell us what parts of the course you think you can use.			
Do you think there are parts you will not be able to use? Why not?			
Other comments?			

Figure 10.4 (Continued)

Training Methods Course Evaluation Form

Please circle the answers that apply and comment wherever you wish. Your careful evaluation and comments will be very helpful in improving the course. Thanks.

What brought you to this course? Volunteered   Required

Describe your training experience and plans

Just getting started   Have never trained   Have trained extensively

Hope to start training soon   Would attend an advanced course next year

Comments:

Were the written materials used for this course:

Interesting   Easy for me to use   Informative   Relevant to topics

Adequate to achieve course goal   Helpful as a take-home reference

Comments:

Was the instructor: Knowledgeable about the topics   Interesting

Good at communication skills   Comfortable in his/her role

Able to make the information understandable   Respectful of trainees

Comments:

Were the training aids (audiovisual, props, etc.): Suitable to topics

Clear and visible   Helpful to understanding   Used well

Comments:

Figure 10.5 Training Methods Course evaluation form designed by student trainers.

Were the teaching methods used by the instructor: Interesting

Suited to the material   Methods you can use   Varied

Comments:

Has this course met the stated goal, and/or your expectations in some areas?

Improve your ability to train   Help you design/develop training materials

Help you design evaluations   Increase your comfort level

Help you deal with difficult situations and people

Help you gain support from people and groups for your training program?

Comments:

What parts of the course were helpful to you?

6 steps of design and development   Training methods   Training materials

Writing objectives (Bloom's Taxonomy, etc.)   Dealing with problems

Hands-on participatory techniques   Getting helpful people involved

Characteristics and actions of effective trainers   Evaluations

Comments:

Do you have suggestions for improving this course? If so, please write them out, as well as any other comments you would like to make. Use the back of the sheet if necessary.

Figure 10.5 (Continued).

VIII. Summary

Please evaluate your training. Try to measure both process outcomes — what goes on in class — and post-training outcomes, especially changes in workplace health and safety. Do this as fairly and accurately as you can. Your primary motivation should be to improve the training, and that requires open and honest inquiry with trainees and a positive attitude toward the evaluation process.

It is important to the future of your training program to show it is effective in accomplishing the goals and objectives you, the trainees, and the financial stakeholder consider important. It is important also to measure financial efficacy. Effective training correlates with reduced accident and injury rates; effective training and improvement in the quality and quantity of production are not incompatible.

chapter eleven

Trainers: born or made?

It is no fun being a poor trainer. Don't lose sight of our goal in being here — to have fun, and to be better trainers. Both are important, and they build on each other. Good trainers — all successful leaders — share certain qualities.

Good trainers are:

- Objective — They lack personal bias.
- Open-minded — They listen to the ideas of trainees.
- Tactful and sensitive — They are careful not to hurt anyone's feelings, and are aware of areas where this might happen.
- Humorous — They have a sense of humor they use to loosen up the classroom and make people comfortable. They are able to laugh at themselves.
- Enthusiastic — They are excited about what they are teaching and happy to see the group that arrives for training.
- Adaptable — They can organize or reorganize things quickly.
- Good listeners — They not only listen to the individuals in the class, they hear what they say.
- Facilitative — They draw others out. Everyone in the class feels included.
- Authoritative — They inspire confidence and trust in training participants.

Some of these characteristics are basic components of an individual's personality, but all of them can be acquired and practiced. It is worth noting that most of them are much more easily attainable if the trainer is comfortable. Comfort follows subject mastery and training practice.

I. Abilities of good trainers

A good trainer — any successful leader — learns to do these things.

- Expresses himself or herself in clear, understandable English (or the language of the class).
- Presents information in a relaxed, conversational manner without reading from notes.
- Demonstrates job procedures accurately, or finds someone who can.
- Uses written materials, audiovisual aids, and a variety of adult training methods to achieve learning objectives.
- Gives clear instructions before the class begins to do workshops, or to practice skills.
- Asks questions to find out if trainees have understood. Doesn't just say, "Any questions?" and quickly go on.
- Evaluates trainees' progress regularly.
- Makes sure each trainee is actively involved in learning.

These abilities increase with practice, and improve with caring. The trainer who is genuinely concerned with training outcomes and with trainees' pleasure in training, as well as health and safety improvements in the workplace, can achieve proficiency in these skills.

II. Special resource people

In certain situations, trainers may want to invite special guests to add their expertise. Caution: Do not let anyone teach your class who is stuffy or boring, or talks down to trainees. You work hard to set a comfortable climate for learning. Don't allow others to pollute the atmosphere.

A. Technical experts

Technical expertise does not necessarily imply good communication skill, which is probably the single most important quality that trainers must possess. Yet there will be times when it makes sense for someone who really knows a job procedure to lead a group of less experienced workers in learning that procedure. A physician might be invited to answer specific health hazard questions, or an engineer to explain how a process works.

Don't overlook other floor-level workers as technical experts. Someone who has worked on a machine for 20 years knows all about it, and refines the operation of the machine by looking at, feeling, hearing, and perhaps even smelling the machine's clues.

B. Training aid experts

Getting expert help to build training aids is a good idea. Carpenters, welders, artists, and multimedia presentation talents are all useful to trainers. Added variety and interest are always welcome in a class.

C. Interesting and motivated speakers

Sometimes it helps to bring in speakers who are funny, especially interesting, or exceptionally motivated. If your training budget does not include payment for professionals, search among your friends and co-workers and arrange some sort of swap for their participation.

Some workers have accident stories to tell, or can describe successes in working out problems. Help them to organize their thoughts ahead of time, and limit their speaking time to 20 minutes if there are no exercises in their presentations.

III. Be aware of the message

You, and all the other people you bring into the training room, send a message to trainees by your actions, words, appearance, body language, and facial expressions. The message is strongest during the first few moments after trainees come into the room. If the message is initially negative, it is harder to send a positive message because you first have to overcome the negative one. If you number public school teachers among your friends, you may have heard the warning given to new teachers from the veterans: "Never smile before Christmas. If things are going well, you may be able to smile by Valentine's Day." This advice is intended to put the new teacher in control of the unruly masses who, it is assumed, will be totally unredeemable if given any slack. For worker trainers, we advise just the opposite. Give trainees a welcoming smile, greet them at the door (you are already in the room, of course, seeing to their creature comforts), and let them know you are glad they came. It's not a sign of weakness to be happy to see them. All this is part of your overall message, repeated several times in this book. "We are all in this together. You know a lot about your job; I know some important things too, and together we can make the workplace safer (or more profitable, or more client-friendly, or whatever is your training task)."

Take a moment to read the one-paragraph descriptions below and state in one sentence the message you think each trainer is sending to the class. Different readers will read different feelings into these descriptions; see what they say to you. All these scenarios are true, and we were there when they happened.

A. Trainers we have known

The safety director of a large organization was asked to make a welcoming speech to a class of members of the organization; the class would be taught by outside instructors. As the class members wandered in and milled around talking to each other and the instructors, the director stood behind the projector screen, effectively hidden from the class. He came out only when it was time to start, spoke very formally, and left.

The safety manager at a large pulp and paper mill was required by upper management to organize mandatory emergency response training for workers on every shift, some of whom did not want to be on the Hazmat team. He contracted with the best training organization he could find; set up training in a large, attractive, well-lit room; arranged for a delicious buffet lunch to be brought in every day; and took the first week-long class himself.

An "expert" (definition: came from out of town with a briefcase) was hired to teach the health effects part of a Hazmat course for truck drivers. He arrived just in time, wearing a suit and tie, and lectured for two hours. He left as soon as his lecture was over.

An "expert" (definition: Ph.D. professor and researcher at a major university) was asked to teach the health effects part of a Hazmat course for chemical plant workers. She arrived in time to talk to students during the break before teaching. She was wearing sweat pants and a T-shirt (as were many of the students), opened the topic by moving around the room responding to trainees' input about chemicals they worked with, then talked about other chemicals and health effects that had not been mentioned.

A trainer was approached on a break by a student who was uncomfortable about the prospect of suiting up in Level A (SCBA and "moon suit") as he had experienced claustrophobia several times in the past. The instructor's response: "You have to do this, or we'll tell your boss you can't get certified."

An experienced trainer refused to update his training methods even after considerable training of his own about effective adult training methods. He frequently turned down the lights so his dull overhead transparencies would be more visible, lecturing for over an hour at a stretch. Often when lecturing, the trainer closed his eyes. So did the class. Many of them went to sleep.

A trainer was asked to fill in for a training group whose regular instructors were at their wit's end trying to please a group of bored and grumpy state regulatory professionals they trained year after year. The new man began by shaking hands with every student and introducing himself. He was filled with enthusiasm, moving around the room as he taught and gathering the personal experiences of the trainees. When he finished, the class gave him a standing ovation.

Two trainers with a very reluctant class ("The boss made me come but I'm not going to cooperate") had a student who suffered discomfort from a bad back. The trainers invited all the class members to get up, move around, and get a soft drink from the back of the room as they wished during training. The student with the bad back was free to lie down on the carpeted floor when her back hurt. The whole class visibly relaxed and the training went well.

A trainer in a course based on an OSHA standard used a thick training manual written by professionals, much of which he himself did not understand. He taught for five days by skimming through the manual, listing the pages trainees "had to know for the test," and spent the rest of the time drilling the class on the definitions of abbreviations and acronyms on the spot, moving down the rows for answers. Most of the students spoke English as a second language and did not read it well. The manual was in English.

Two trainers went to west Texas to teach a week-long class, and found on the first morning that five trainees spoke no English. Half the class was white, highly educated environmental professionals; the others were Mexican-American and African-American laborers. The trainers hired an interpreter, mixed the trainees up in a variety of groupings for different participatory exercises, and had a "graduation party" at a local park in the desert. Everyone passed the course, and everyone came to the party.

B. Great trainers we have known

Just as we can always find someone richer, thinner, or better looking than we are, there are always trainers who are better. They are gifts — take advantage of them. If you are fortunate enough to work occasionally with a great trainer, use that person as a model. If his or her methods mesh with your training philosophy and personality, try them.

We have had the happy experience of getting to know, and convincing to work with us when they have the time (they are both in considerable demand from training organizations), two excellent trainers who always brighten our days, and certainly brighten up our classrooms. What follows is an attempt to dissect and label their best qualities in order to improve our own training skills. You will find physical descriptions included here: Do they matter? Who knows, but perhaps appearance and dress, although not as important as attitude and competence, influence trainees' first impressions and set the tone.

Robert is a handsome safety professional, stylish dresser — generally no tie, but nice slacks and shirt — who makes an initial impression of friendliness and attentiveness that grows as trainees spend time with him. He greets each trainee personally, and it always seems to us that by the time class starts he has found something in common with each one of them. He personalizes his training by using individual's names and seeking input from each person. His computer-based presentations are not outstanding in design — they contain no arresting sounds or graphics — and the number of slide changes is minimal so that he can pay attention to the trainees and not the mechanics. By remaining flexible and adapting the lesson to the class members, Robert manages somehow to elevate each trainee's spirits and convince each one that life can be better. Jaded trainees almost battle each other to participate actively in his sessions. Robert is the trainer who received the standing ovation from the previously very bored class described above.

Robert tells about his first training experience. "It was terrible. I really bombed. They wouldn't invite me back, so I began asking friends to please let me go somewhere and speak, do some training, practice until I got better at it." Recently, Robert has been the invited motivational speaker at several conferences.

Figure 11.1 Sam does rope tricks in the classroom...

Sam is a Southern good ol'boy, an athlete and ex-baseball player who often trains in jeans, with an Alabama accent so strong that the Yankees in the class have to listen closely to understand him. His bag of tricks includes a few actual magic tricks, used when teaching knots to a confined space rescue class (Figure 11.1), small but alarming fires and explosions (for demonstrating the properties of hazardous materials), and Tinker Toy sets for building a "dilemmasaurus" as an exercise in improving communication during Hazmat incidents. Video footage of real incidents is reduced down to the five most hair-raising minutes and played to grab the audience. A tape recently used at the start of a rope rescue class shows a firefighter and the victim he has rescued from the window ledge of a burning building; they are spinning at the end of a manila rope 13 stories off the ground.

Sam's best attention-getter is that he knows his stuff! His experience with Hazmat response and high-angle rescue is extensive, and it shows when he trains (Figure 11.2). If he teaches for an hour in a three-day course, Sam is the one they single out on the course evaluation. Sam's advice for trainers is "It's gotta be fun. If you hate teaching it, they'll hate learning it."

Common traits shared by these two trainers include a ready smile, voice contact with each trainee including the use of names, excellent knowledge of the training topic, and movement about the room to include everyone.

Figure 11.2 ...then takes his expertise outdoors.

IV. Setting the scene for learning

Trainees learn much more easily when they are comfortable, and the trainer holds the key to much of their comfort. A comfortable training setting depends on physical conditions as well as the psychological climate. Do as much as you can to improve the classroom setup, the equipment you use, and the physical climate.

A. Classroom setup

Obviously, you will have to play the classroom you are dealt. The following are suggestions we have found to be of help.

- Trainees need tables to work on. The next-best thing, but a poor second, is a group of desk-type chairs with attached writing arms. Rows of chairs are not adequate. Auditorium-style seating, where chairs are fixed in place and cannot be moved around for group work, is unacceptable.
- Arrange the tables and/or chairs so that everyone is a part of the group, and everyone has equal access to your attention. This means the best arrangement is a **U**, with the trainer at the open end.

- If a U is not possible and you have to arrange rows, leave an aisle down the middle so you can move as you talk. This way, you can include everyone.
- Be sure that for small-group workshops the chairs can be turned so that the members of the group face each other across a table. Each group member needs to be able to see all the faces in the group.

B. Equipment

Chairs and tables are a must. In addition, you will need the following.

- Some sort of writing surface and writing implement for the trainer, perhaps one or more of the following:
  - chalk board and chalk
  - white board and erasable markers
  - flip chart, with stand and markers
  - large roll of kraft or white paper, tape, and markers
- Special aids for certain classes as noted in the lesson plan for that class:
  - video player and monitor or projector, and screen
  - prepared posters
  - laminated maps or drawings for scenarios or site planning
  - overhead and/or slide projector, with screen

C. Comfortable Climate

People need to be physically comfortable in order to pay attention. Physical discomfort distracts from your lesson. Here are several suggestions to ensure physical comfort.

- Set the lights for bright, and the thermostat for comfort. If the thermostat is in the training room, point it out and invite trainees to change it if they get uncomfortable.
- If you can possibly do so, arrange for coffee, soft drinks, water, and snacks. Invite trainees to graze whenever they feel the urge.
- Let the class know at the beginning where the bathroom is. Assure them you will take frequent breaks, and they don't need a hall pass to leave between breaks.
- Some smokers cannot keep their minds on you for long periods of smoking deprivation. Smoking in class is a health hazard, and is no longer socially acceptable in most settings, so if your class includes truly addicted smokers you will probably want to take more frequent breaks.

D. Psychological comfort

A number of techniques can be used before and during training to help trainees relax and get the most out of training.

1. Communicate the Ground Rules

Ask the group to set guidelines for training discussions and what can happen with information that is revealed. Although you cannot force people to do so, ask for confidentiality of the information and feelings that are discussed.

One professional trainer suggests you ask the group to record on flip-charts the answers to these questions.

- What can I do to make this session successful for you?
- What must I avoid doing to make this session successful for you?
- What can you do to make this session successful?
- What must you avoid doing in order to make this session successful?

These are discussed and posted in the training room.*

2. Encourage interaction

There are no dumb comments or questions. On the other hand, if a trainee tells the class a "fact" that you know is incorrect, you will want to find a way to get the correct information to the class without making him look stupid. You might counteract the information with a different interpretation backed up by a research group or government agency, or by looking it up in one of the references you have in class.

3. Stress flow of information

Learning is not a one-way flow in which information moves from instructor to trainee. In a good class, everyone involved brings knowledge, experience, and insights into the classroom. Encourage questions, comments, and the sharing of relevant experiences that will allow everyone present, including the instructor, to learn from each other. Questions and comments provide valuable feedback, allowing the trainer to maintain a "real world" focus. Design group work into the schedule so interaction is encouraged and directed toward a stated goal. This allows an exchange of ideas without the problems associated with too many individuals voicing their comments in the general session.

4. Try things out

Try new things in the group environment, and don't be afraid to make mistakes. You are here to facilitate learning, not to show off your expertise. Some people feel using untried methods is analogous to cooking a new recipe for guests, and point out the likelihood of failure. We insist on trying and using new worksheets, scenarios, and problems before setting them before the class. Certain things, however, can be tried only in the setting where they will be used. Here's an example.

* From the December 1998 issue of *Creative Training Techniques*, Lakewood Publications, 50 South Ninth St., Minneapolis, MN 55402. 1-800-328-4329. All rights reserved. Reprinted with permission.

A training methods class for paper industry workers who wanted to train other workers had 41 participants. We were there for three days, and wanted to use the Fun Facts icebreaker, described in Chapter 5, on the second day. Since 41 names on that page would be far too many, we decided to try Group Fun Facts that were common to all members of the practice training teams already set up for the final day's team teaching. On completion, the instructor stated she didn't think it worked well, and asked the opinion of the class. They thought it was a great way for the members of the teams to get to know each other better, as they had to ferret out individual experiences to write a fact the entire team shared.

V. Dealing with problems

Stuff happens. Trainers deal with it. The power fails, the projector bulb burns out, trainees get into arguments, and a trainee starts talking and won't shut up. No trainer is born knowing how to fix a balky remote control, or handle an irate worman who is mad at everybody she works with. The longer you train, the more "interesting" situations you will find yourself in. Consider each an adventure. You know the definition of an adventure: An event that is more fun to talk about later at the bar than it was while you were living through it.

A. Physical problems

Some facility and equipment problems can be anticipated and prevented, and others cannot. In advance of training, you should do the following checks.

- Check your equipment to be sure you have it all and it works.
- Bring extra projector bulbs, fuses, and a plug adapter.
- Practice with new equipment before you go into a class.
- Check the training room or area ahead of time.

Things will still go wrong, and there are no universal rules for handling them. Every experienced trainer has war stories. All of these adventures happened to one of the authors.

1. One member of a training duo got thrown in jail in the middle of the night with the equipment van keys in his pocket. Class was 30 miles away.
2. The slide projector in a conference center was being operated by the remote control used by the trainer in the next room across the temporary wall. Slides seemed to change by themselves, controlled by a ghostly hand.

3. The boxes of manuals and overhead transparencies got lost in shipment and didn't arrive until the second day of a week-long course.
4. The laptop computer refused to work.

There you are. Handle it. As you gain training experience, this gets easier.
As you become more confident and relaxed, more options are possible, and
you will find them. Being able to tap dance is a useful skill, literally as well
as figuratively. On a more practical note, here's how we handled the four
adventures described above.

1. The incarcerated trainer's partner phoned the industrial facility where
   the class was taking place and gained an hour's delay. She walked to
   a car rental agency, rented a car, drove to the training site, and taught
   both partners' material. She explained to her partner she really didn't
   care if he was angry that she did not show adequate sympathy for
   his plight, and recommended he not be found drunk and barefoot in
   the hotel parking lot again. He paid for the rental car, and she didn't
   tell on him.
2. That entire room was a mess. It was dark; there were no tables for
   seminar participants to put their materials on; the screen was far
   behind a huge podium with a high railing; and the slides changed
   themselves. Since the audience consisted of trainers, they were invited
   to critique the room, equipment, and setting, and then change it. The
   remote slide changer was disabled. Following the room rearrangement, the seminar went on as scheduled.
3. Fortunately, when training at a distance, all prepared trainers carry
   on their persons the training aids required to teach the first day's
   topics. These trainers were prepared. Although they taught without
   student manuals, they paid for student copies of necessary outlines
   and worksheets. One wise old trainer had given them this advice
   during their early years: "Always be sure you have a half day's
   training materials, a plug adapter, and a corkscrew."
4. The computer. Oh, well. Even the best trainers can't fix everything.

B. Human problems*

People problems are harder to deal with than equipment problems, because
trainers have less control over them. In some cases, you are training people
who are resistant to your lessons, and in others the problem is the personality
of a trainee, or the way he wants to express himself.

---

* Clip art is used in this section of Chapter 11 to create a mood. It also serves to break up the
page, a tactic recommended for writing interest-capturing, user-friendly training materials. All
but one of the actual images are colorful, and we use them on overhead transparencies and in
computer-based presentations. Be cautious of copyright restrictions when using clip art in
published manuals; generally, they can be used in-house with no special permissions.

1. Dealing with resistance

One of the most common human problems trainers encounter is resistance from the trainees. Here are some of the things trainees resist.

a. The training setting. Some people respond negatively to a class-room setting itself — they don't like to be "in school." By now you know how to make people feel comfortable. Resistance to the schoolroom setting probably comes from past experience, where students were bored or felt unsuccessful. Fear of being put on the spot for answers is another factor. Participatory training deals with all three of these worries. Participating trainees are not bored; they can be successful because success depends on doing real things with the group's help, and you never put them on the spot for a verbal answer.

b. Attending the session. Trainees may resist simply being there, especially if they've been forced to attend training. You really can't change the fact they have to be there, but you can help them figure out the session won't be too painful. Show your respect for trainees and the time they feel they are wasting. Use participatory training methods.

It may be helpful to talk about this as you begin class. Ask how many people volunteered to come and how many were sent by their boss. By letting them know that it's okay to be less than happy about being there, you acknowledge people's feelings and show than that they are not going to be judged for feeling that way. Your attitude can be, "We're all in this together, so we may as well do something while we are here," but don't overdo the happy talk; you'll only irritate them.

c. Authority. Some people resist any form of authority, including the person standing up front in the position of trainer. You are being blamed for something that is not your fault, but it doesn't help to say that. Use concrete visual clues. Don't stand at the front. Don't stand behind a podium. Don't let your mouth or your body language say, "I'm in charge.

Do it or else." Remember, you are here to figure out the best way to accomplish something, with the help of the group.

d. Change. Many people don't like change. Older workers especially are likely to resist new methods. Leave them alone until they realize you are not forcing any changes on them, but that the group is looking for ways to do things a little differently with input from all. Group activities where everyone works on policies or solutions to safety problems will allow input into the changes, and a feeling of ownership of the new ideas.

e. Anxiety. Resistance may be due to anxiety. Being asked to learn or change behavior, or practice a new skill, makes some workers anxious. Here we are probably dealing with fear of failure, or fear of the inability to "get it right." Never announce you are here to change behaviors. Knowledge, new skills, and different behaviors will evolve out of the group process if you guide the process and provide the necessary information. Be empathetic about the anxiety some participants feel about physical tasks. Reassurance from the trainer can help trainees overcome most fears (Figure 11.3).

f. Recognizing resistance: what to do? Recognizing resistance is possible through the clues provided by trainees, even if they don't openly express resistance. They glare at the trainer; they sit back in the chair with arms folded over their chests; they keep their dark glasses on. Since the individual is already in a hostile state, it makes sense not to give him (or her) a target. Don't ignore him, but ignore his behavior unless he makes it impossible for you to do so. It's probably not personal; after all, he hardly knows you. Go ahead with your best weapon — interesting, involving activities. He'll probably get over it. If not, he is likely to be quietly uncooperative.

The resisting trainee may decide not to participate. That's his choice. Attempting to cajole or force him to actively participate will only call attention to his attitude and reinforce it, so leave him alone if he doesn't respond

Figure 11.3 Provide reassurance and support to trainees anxious about suiting up.

to your initial efforts. Count him into a group just as you do everyone else, but let the group deal with him if he refuses to play by their rules. As long as the resistant trainee is not creating a disturbance, ignore him. After all, it is his loss and not yours if he refuses to take part in activities that are designed to increase his knowledge and skill. Remind yourself of the advice about why you shouldn't try to teach a pig to sing: it won't work; and it annoys the pig.

2. Dealing with problems of expression

Some trainees express their ideas and feelings in ways that are not conducive to allowing the class to move forward in the direction the trainer wants it to go. People talk too much or not enough; or they argue with the trainer, complain about the company, or say the training is a waste of time because nothing is going to change as a result.

Trainees' means of expressions are a problem only if they affect the flow of the class, waste participants' time, or cloud the classroom with negativity. If you can complete the session without too much series disruption, you have done a good job. Figure 11.4 illustrates a truism to keep in mind about dealing with problems of expression. Hang it on your wall, and refer to it often.

> When you place a group of people with common needs, ideas,
> or problems in a room together,
> they are going to talk about those needs, ideas, or problems.
> Give them a structured time to do this,
> or they will do it on your teaching time.

Figure 11.4 Hang this trainer's mantra on your wall, and refer to it often.

Early in the class, design into the agenda a small-group exercise where the groups are free to brainstorm about a topic you have chosen. Monitor their progress toward answers, and keep them within the time you have allotted for the activity, but don't jump into their group if you notice they are talking about common problems or gripes other than the assigned ones. By the time report-backs are called for, the group members will have blown off some steam and expressed the gripe they brought to class.

a. Control expression early.   Many trainers start a class with the request to "Go around the room and let everybody introduce themselves." We have seen this turn into a series of speeches describing the problems that trainees are having, with the lame connection that they hope the class will help with the problem. Really, they just want to tell their tales. It may take 30 people over an hour to tell all the tales, and you probably don't want to give up that much time to unstructured and unproductive stories.

Do this instead. Ask each person to say his or her name and department (or other identifier) and then to state one thing he or she wants to get out of this training. Reduce these to a few words and write them on the flip chart. If an individual starts on a long background story explaining why he wants the class to provide certain knowledge, gently interrupt by saying, "So what you'd like to get out of this class is information about respirator selection" (or whatever), write it down, say "Good. We'll cover that," and go on to the next person. Or skip the formal introductions — do an icebreaker instead.

b. Controlling the runaway talker.   A trainee talks too much. Either he has a comment for every topic and at every opportunity, or she rambles and won't bring her comments to an end. Try these physical and verbal actions to slow the runaway talker.

When you ask for comments or ideas, look at someone else. Physically move away from a repeat talker, place yourself where you can't see his hand go up or his mouth move, and request a comment from a different area of the room.

Ramblers are harder to stop. You sometimes have to interrupt them to move on, and that is not polite. If she stops for breath, immediately affirm her contribution by a positive response ("Yes!" or "Thank you!") and quickly move your attention to someone else by asking, "Do you have an idea on this?" If you don't want another comment, take advantage of her breathing break to summarize her points and yours. Then move on to your next topic.

If you are unfortunate enough to get a truly obnoxious rambler that causes everyone else to groan when he starts to talk, take him aside at the next break and ask him to give others with less experience a chance. Before that, there is little you can do unless you are willing to put out your palm like a crossing guard and simply say, "Stop." If you have tried every other tactic and the class is suffering, you may have to do this.

c. Trainees who do not talk. A trainee talks too little, and does not contribute to the discussion. This is really not a problem. Not talking doesn't mean the trainee is not participating. Maybe he's just not much of a talker. He probably will contribute in small group exercises, so schedule some.

d. The chronic complainer. The trainee wants to complain about his life, work, boss, spouse, or co-workers on your class time. Start by saying something like, "Yeah, that's a problem," and attempt to go on. If he persists, tell him that isn't what the class gathered to deal with so you need to go on with the agenda. If the problem is one you have gathered to deal with, express the hope the group will be able to help him with the solution. Repeat as needed.

If the complaining trainee is with you for several days, you can begin to bend his comments toward ideas about what he might do to correct the problem. When the third repeat of the same complaint starts, interrupt and say, "Yes, but we're here for solutions. What might you (with voice emphasis on the you) do to change that situation?" At best, the trainee may come up with some good ideas. This trainee will not stop these annoying outbursts at this point, and you will have to be vigilant. Soon, you are holding up a hand as he begins and saying firmly, "Solutions!" At worst, he sees that you are not going to let him get by with repeating the same complaint over and over without giving some thought to doing something about it. At best, he redirects his thoughts toward taking action to solve the problem. This works, we promise. Try it.

e. The arguer. Trainees argue with each other. This may bring out some interesting ideas if the argument stays on a friendly, informative basis. Let it go for a while until it strays off the subject, threatens to go on forever without some sort of agreement, or becomes personal.

The authors once stood by as members of two public agencies argued about the relative merits of their actions in a chemical spill. It was a productive argument until it got personal and heated, at which time the class asked the trainers not to interfere because they wanted to see the firefighter punch the cop. We intervened and a fight was averted, much to the disappointment of the class.

A trainee argues with you. If you have not presented yourself as an all-knowing expert, you can handle this gracefully by summarizing the points expressed by both sides. If you are sure of your information, you may want to ask a class member to consult a particular reference, or another to call on his expertise in the field to provide a framework in which to consider the conflicting viewpoints.

The way you handle this will depend a great deal on who and what you are. A grandmotherly Southern woman might say, "Bless your heart, you may be right. Let's look that up." A laid-back industrial hygienist might pleasantly repeat something one of his graduate school professors said, while a young trainer who is not sure of his technological expertise or his facts will probably stiffen his neck and his manner and confront the trainee, to the detriment of everyone involved.

One experienced mediator uses this technique. "Hmmm," he says, as he strokes his chin and looks at the ceiling as if he is considering the conflicting opinion. After a moment's quiet, he continues right on with what he was planning to do without acknowledging the opposing comment. The group is surprised, and goes along with him.

And don't forget the Donna Reed method. Remember TV's Donna Reed? As she vacuumed the carpet in her shirtwaist dress and high heels, her hubby and children reported the crises of the day. Donna would tilt her head, look at them attentively, and say "Oh." And that was it. Resume vacuuming. This tactic works well with all kinds of people who want to stir you into a reaction you do not care to make. It works also on trainees. And spouses. And friends.

f.   Extreme negativity. A class member is a negative that she doesn't believe anything good will come of the training, and expresses herself at every point. Try to leave this person out of the discussion if you can. If her attitude changes, it will not be due to anything you said. Your best recourse with this person is to form a consensus with the rest of the group, and hope she will feel outnumbered and come around.

If your "wet blanket" insists on inserting negative comments into the discussion, ask her to suggest an alternative. Keep doing that until she will no longer enjoy her role because she knows you are going to continue to request she think of a positive aspect.

C. Dealing with real problems people bring to class

Remember Barney? He's the man in Chapter 1 who took over a class in an angry, somewhat drunken state to complain about being laid off. Because he was sitting at the front, facing the class, the trainer was able after a few minutes to move down the center aisle and direct some questions to class members to regain their attention. Barney did not immediately shut up, and was not lucid enough to pick up on the clues the trainer was giving him (the strongest was that her back was turned to him). The trainer was trying to accomplish two things: quiet Barney down, but don't make him feel humiliated. Everyone there wanted him to stop talking so they could get on with the class, but all were sympathetic with his plight. If a trainer stoops to embarrassing a trainee, even an obnoxious one, she loses the confidence of the group. Barney soon wound down, and the trainer was able to continue with the activities. A break was called, and Barney's fiends took him outside for a smoke break from which he did not return.

Many real workplace problems can be solved by the group, as there may be another worker in the class who has faced the same sort of difficulty. A good trainer utilizes every opportunity to get group input into solving problems. If the discussion shows signs of being too long, and will divert the class from its real mission, end it and later put the people who share the problem and its solutions in a smaller activity group, with the assignment to report back later.

D. Setting the tone and spreading the attitude

If trainees are with you all day or over a period of several days, they will pick up on your attitude toward what the group is trying to do. Basically, your bottom line is this: We can discuss problems in the context of solving them, but we are here to seek solutions, not repeatedly complain about existing conditions we may not be able to change, especially if all we do about them is gripe to each other.

Every time your turn a rambler or a complainer around by focusing on the main point, or by requesting an idea leading toward a solution, you reinforce the bottom line. Trainees are perceptive. Never lose sight of the bottom line, or lose your "We're all in this together, so let's go for some answers" attitude.

E. Trainers and responsibility

When something goes wrong in class, is it the trainer's fault? How much responsibility does a trainer take for what other people do during a training session? Certainly, if someone is injured because the trainer did not reasonably anticipate a danger, moral and legal liability may ensue. We are not prepared to take on that issue here. Danger aside, events occasionally conspire to leave trainers shaking their heads and wishing they had anticipated, or intervened in, classroom difficulties. The purpose of relating these two stories is to provoke the reader to plan ahead. It will happen to you when you least expect it.

At the close of a three-day course, a trainee asked to say a few words to the class before they left the beachside resort where they were meeting. The trainer agreed; and the man launched into a hellfire and damnation religious sermon, during the course of which he castigated class members for their off-time behavior, including the disappearance of an empty wallet he had planted on a bench outside the classroom.

In another course, a group spokesperson was designated to give a short safety lesson to the class. He began to draw, line by line, an explicit female form on an overhead transparency. Several class members, as well as one of the instructors, were women, and were visibly uncomfortable. The last few lines drawn turned the naked women into a long-eared dog wearing hearing protection, but not until well into the lesson.

Both these situations happened to us. Both caught us so off-guard that we did not handle them well. Upon reflection, with input from our complete training staff, we agreed on what we should have done. The stories are repeated here as a warning: Think about them now, not when it happens to you. Your problem situations will be different from these, but if you train long enough, they will occur. What will you do? How much responsibility do trainers have for what goes on in their classes?

F. Training trainers

If you are a trainer of trainers, you can help them train effectively by helping them become knowledgeable and comfortable. Support them with technical seminars, adequate financial resources, good materials, a competent staff, and plenty of practice time. Be clear and specific about the things they are doing well, and equally clear and specific in discussing areas that can be better. Give trainers-in-training measurable objectives to attain, just as you would for any group of students.

Teach your trainers how to use all the equipment they will need, and provide plenty of time for them to practice using it. Confidence in their ability to deal with machines and gadgets will help to remove one source of nervousness. When they train for the first time, be sure they have detailed lesson plans from which to work. Also, send them in pairs so they can help each other with little things like light switches, which are always located at the opposite end of the room from where the trainer needs to hit the PLAY button on the VCR. When a trainer's mind goes blank on a question, it helps to have another brain in the back of the room to give the response.

Unless you know them well, make no assumptions about what your trainers-in-training already know about teaching. Starting from scratch will not hurt the ones who already know something — they can use a review.

The greatest problem for new trainers (and some old ones) is the fear of being in front of a class. Others have suggested fear is lessened by specific methods: look over their heads; look someone in the eye; visualize your audience naked. We suggest simply that doing it is the best way to learn to do it. Practice, experience, more practice, and more experience are the only way to lessen fear. The more times a trainer does it well (some trainers define that as "without major disaster"), the easier it is to train again. At some point, leading a class becomes pleasurable without any pain at all.

Chapter 13 offers activities for training trainers. All have been tested in train-the-trainer programs.

VI. Summary

This chapter has been about comfort; the trainer's comfort, and that of his or her trainees. The trainer has the responsibility to ensure that trainees are comfortable, both physically and mentally. It takes extra time, arriving early, moving the furniture, and calling the person responsible for fixing the broken lights, but it is worth it. When class members are comfortable, they are kinder and more agreeable, and the instructor's life is easier.

Trainers are not born — they're made. You can make yourself into an effective trainer with experience and practice. The happiest trainers are those who are confident in their skills and expertise in both subject matter and training abilities. The more you train, the better you get if you are willing to work on your skills and expertise.

When you have reached a point where you are really confident and want to add the final polish to your training skills, try this suggestion from an experienced trainer: it's important for trainers to receive feedback on how the group perceives them as trainers. After the first break on the first day, invite participants to draw a cartoon of you on a separate flipchart set in an easy-to-see corner of the training room. The cartoon undergoes modifications at subsequent breaks as the session progresses. Carefully examine the cartoon to detect any tell-tale negative signals, and make corrections as needed in your methods.* The result is invariably hilarious, and gives everyone a laugh — including the trainer. If we ever feel secure enough to try it, we'll let you know how it worked.

Remember, though, however you go about it: The best training is a partnership between the instructor and the class participants. The more skills you bring to the partnership, the more likely you are to be able to train effectively while everyone, including you, has fun.

References

1. Pike, B., Ed., *Creative Training Techniques*, 11:12(3), Lakewood Publications, Minneapolis MN, 1998.

* From the December 1998 issue of *Creative Training Techniques*, Lakewood Publications, 50 South Ninth St., Minneapolis, MN 55402. 1-800-328-4329. All rights reserved. Reprinted with permission.

chapter twelve

Training program support

A successful training program has the support of all the people it serves: workers, supervisors, lower and upper management, and special-interest groups such as unions and health professionals. By circumstance, trainers often are dropped into situations where support is lacking and must be gained through painstaking undercover operations. Let us simplify the definition of undercover operations: find out what people want and give it to them.

I. Giving workers what they want

Once upon a time, a trainer turned on a video camera and asked a group of workers to give advice to a plant safety director/trainer who had just come on the job. The trainer soon turned off the camera because the advice was repetitive. Every worker gave a similar answer.

Question: What should the new safety professional do first? Answers: "Get out of the office, get out on the floor, and find out what we do out there."

Question: What's your safety training program like? Answers: "They have us watch a boring video for 30 minutes, then sign a paper saying we are trained," and "The last guy they hired to train showed up in a tie. He had no idea what it takes to do my job."

Whether these quotes reflect your experience or not, they are real comments from real workers who came from a variety of workplaces. They were especially hard on young safety managers fresh out of college, who had no experience on the shop floor, and on "softies" of any age who had never worked physically hard for a living.

No matter what your age or experience, the people you are hired to train will cooperate with you if you approach them as equal partners in a safer workplace. The best way to find out how to train workers well is to ask the workers themselves what they need, what they want, and what they like. The best way to find out what works and doesn't work in training is to be willing to go out on a limb and try their suggestions. As long as you do not present yourself as an expert on how workers should do their jobs, they will work willingly with you to make the training program effective.

Asking the target group is not the only way to find out what training workers need. You must also turn to OSHA, MSHA, and EPA regulations; accident and injury logs; workers' compensation records; and any other sources that provide clues to safety training deficiencies. A safety manager in a paper company where a Behavior Based Safety program is in effect found that records documenting unsafe behaviors generally pointed back to training deficiencies when compiled and examined. Differences between workplace settings and individual differences among workers also direct and shape the training plan.

A. The "average healthy worker"

You may have heard of OSHA's "average healthy worker" on whom exposure limits are based, as in "the level at which the average, healthy worker will not experience adverse health effects." This phrase, although not found in OSHA documents or American Conference of Governmental Industrial Hygienists (ACGIH) literature, is widely used and probably is an expression of the ACGIH assertion that "nearly all workers may be repeatedly exposed day after day without adverse health effects" to ambient air with specified concentrations of air contaminants.

Experience with workers demonstrates that no worker can be held up as the average worker. Since every human being is genetically unique, it follows that each worker is at least slightly different from the others in every way, including all the factors that are important in safe work and safety training.

B. What do workers want?

You are reading this book because you want to provide effective training to a group of workers in order to protect them from workplace harm. Protection will result from their choices and actions, which will be the outcomes of understanding and doing what you advise. The most effective training will give workers what they want; they are then most likely to understand, remember, and follow your lessons for safety and health.

What do workers want? In several different contexts, workers have been asked to rank these or similar factors by order of importance to them. Rank the list to show the order you think reflects what the workers in your training population want.

- Good working conditions
- Feeling "in" on things
- Tactful disciplining
- Full appreciation for work done
- Management's loyalty to workers

- Good wages
- Promotion and growth in the company
- An understanding of personal problems
- Job security
- Interesting work

After you have ranked the list, look below to find out how several different groups of workers have ranked the factors. In order of importance, here's what they said they wanted.

1. Full appreciation for work done
2. Feeling "in" on things
3. An understanding of personal problems
4. Job security
5. Good wages
6. Interesting work
7. Promotion and growth in the company
8. Management's loyalty
9. Good working conditions
10. Tactful disciplining

How did your list compare with the rankings of workers? How well do you understand what workers want? As a trainer, how complete is your knowledge of the people you train? Study the rankings from the point of view of a training program planner. Look at the items at the top of the list. How can trainers provide them? By including workers in the selection, planning, implementation and evaluation of training.

C. What else do we need to know about workers?

We know not all workers are alike, and not one of them is average. We know what they want, generally, from the company. If we are to utilize fully the expertise and abilities of workers in safety and health improvement, and in our training programs wherever possible, we need to know their characteristics and capabilities. In Chapter 5, we examined common characteristics of many workers. What are their attitudes toward work, and how readily do they take the responsibility for solving safety problems?

Some very old management theory held that there were basically two types of managers; Theory X and Theory Y managers. The two types had very different ideas about workers, as you can see in Table 12.1.

Today, most managers acknowledge that Theory Y better describes workers who are, after all, just a collection of human beings very similar to the rest of us. The characteristics described by Theory Y all can be useful to trainers as they plan, design, and carry out training programs with the cooperation of the workers who will be involved.

Table 12.1 Theory X and Theory Y Managers

Theory X	Theory Y
People hate to work	Work is pleasant to most people
Workers don't want responsibility, and they want to be told what to do	Workers prefer self control, and achieve more when they have it
Workers have little capacity for solving problems	Most workers can help solve organizational problems
Workers are motivated only by physical and safety needs	Workers are also motivated by social and esteem factors, and the wish to do a good job for the sake of a job well done
Most workers must be closely controlled to achieve objectives	Workers can be self-directed and creative if they are motivated

II. Giving bosses what they want

The company is in business to make money. One way to make money is to avoid spending it on accidents and injuries, workers' compensation, and lawsuits. Although there is often a perceived conflict between safety and production, processes and operating procedures that are developed through careful analysis can generate increased product, improved quality, and decreased accident and injury rates.

Bosses are human, and they care about workers. No one wants to see a worker killed, injured, or exposed to chemicals. Bosses want workers to be trained to work efficiently and safely, and to demonstrate their understanding that they have a stake in the bottom line (profit). Bosses want employees who don't make waves — who effectively go about their duties, making suggestions, solving problems, and dealing with ordinary situations that come up from day to day. Bosses want respect and appreciation for their decisions, including those they make about training.

III. What do trainers want?

Since you are probably not the "average healthy trainer," you will have to answer this question for yourself. Trainers surveyed by the authors answered as follows, without ranking the list.

- Enough time and financial support to do a good job
- Appreciation from the workers they train
- Appreciation from the people they report to
- Freedom to make decisions, try new things, take risks
- Control over their own programs
- Reasonable work load and flexible work hours
- Job security; good pay and benefits

- Proactive, rather than reactive, involvement in program development; in other words, to be part of the planning (just like workers want)

About half of the items on trainers' wish lists are under their control, but may need a jump-start and persistent work to get them accomplished and institutionalized. The suggestions in this book, including the ones in this chapter, are designed to help trainers get what they want by giving other people, especially trainees, what they want.

IV. Good training is cooperative training

The wishes of trainers, trainees, and bosses are compatible. We can all have what we want. It is not too much to expect that workers, trained in an effective program they helped to plan and present, appreciate and are in agreement with what trainers offer, and bosses are so happy about the results they give trainers free rein and full support.

The most effective worker training depends heavily on resources from workers, trainers, and managers. A great deal has been written about the importance of involvement by everyone in a training program. Management styles, safety programs, and labor/management relations are moving in the direction of more cooperation.

The more people at every level who have ownership of the training program through being part of the planning and delivery process, the more these people will support the program and have a stake in its success. Research by labor educators verifies the hypothesis that the more control and ownership workers have in safety training, the more effective is the training.[1-3]

A. Recruiting cooperation in the training program

Running a training program is a big job, as you know if you are already doing it. You work extremely hard to put on a great course, and still people complain: It costs too much; they don't have the time; you're requiring them to waste time on things they already know; training is boring; they know more about how to do their job than you do — and we could go on and on.

There are several very good reasons to recruit help. For one thing, people who have been complaining have a chance to change the things they are complaining about and complaints will decline. Better yet, you will have included people with a range of expertise who can add to the quality of the training program. These two factors are very important. Read them again.

If you are new at your job and have not learned the politics of the organizations you will be dealing with, be very careful. The hierarchies in companies, worker groups, and labor unions can't be ignored. Find an experienced person at each level, if possible, one who puts the interest of his or her group above or at least equal to a personal agenda, and ask these people frankly how you should proceed. Most people will respond favorably to being asked for input. A false step early, even if taken in ignorance and not from malice, is hard to overcome and can interfere for a long time with the cooperative feelings you are trying to engender.

B. People who can help

There is a wealth of knowledge and ideas among people in your company. Upper management, supervisors, labor unions, safety committees, unorganized workers, the plant nurse, the plant fire brigade or Hazmat team — this is only a partial list of those with something to contribute to your training program.

1. Upper management

You need upper management's help to get money, time, and credibility for training. The most successful training programs are those that are supported, attended, and followed by management all the way to the top. From the perspective of workers in the plant, managers who do not participate in safety training and who are observed not complying with the things taught in the training program undermine even the best safety training.

The most important person in the upper management hierarchy is the one who has the power to make decisions about spending money. The most important manager to impart credibility to your industrial training program is the plant manager. Probably they are one and the same. To paraphrase the T-shirt, "If the manager ain't happy, ain't nobody happy."

a. The money factor. The cost effectiveness of training is probably your best tool to pry out the money to support adequate training. Your argument can best be supported with facts, so take the time to locate accident and injury logs; approximate costs of workers compensation, lawsuits, and OSHA fines; and include supporting comments from your loss control and insurance people. If you are a worker who serves as a peer trainer, show the company figures that prove how much money you can save by replacing outside "experts" and contract training providers. Demonstrating your qualifications to train will be an important factor, so plan ahead for that. Take a training methods course if you can, or be able to show teaching experience in another setting. Gain certification in the subjects you want to teach.

b. The quality factor. You will also need to provide proof of the quality of the training you can deliver. There are several concerns here: The knowledge and background of the people who design and write the training; your credentials as a trainer; whether the training is compliant with OSHA and other standards; and whether you have included provisions to document workers' participation and success in training. Documentation of training is a big concern of management for loss control and regulatory reasons, and must include trainee names, dates, content, and the qualifications of the trainers.

c. *The "buy-in" factor.* Management must buy into your training and support it philosophically as well as financially. Managers can extend the safety and health influences of training in several ways.

Before a training session, managers should let workers know about the benefits they will receive from the class. Reinforcement is critical to the adult learning process. Invite top managers to attend the opening session and spend five minutes telling participants about its value and their support.

As soon as the workers return to work from training, floor managers should immediately take the time to begin the reinforcement process. Ask about the session, what was learned, and how workers see it being applied in their area. Ask about obstacles to implementation that workers anticipate, and get their suggestions for overcoming them.

If the workers you train like the training and are willing to say so (which you ask them to do when you design course evaluations), management will be happy. If workers like your training they are more likely to take it to heart. Production may improve, and safety certainly will. Be sure to design into your planning, training, and evaluation the means to document trainee feedback and your response. If you can also document a reduction in accidents and injuries, work-related illnesses, and lost time and tie the reductions to training, management will be ecstatic.

2. Safety committees

If you have access to safety committees, use them. There are more joint labor/management committees set up to deal with health and safety issues than for all other issues combined. Assume members of the safety committee are concerned about safety, and seek their help in designing safety training.

A good safety committee has operating guidelines that you can tap into, such as setting topics in advance of the meeting and sticking to the topics when they meet. Get your topic on the agenda, ask them to investigate the issue of training with workers they represent, and really listen to them when they report back. If it seems feasible, ask the safety committee to set up a subcommittee on training and include them in your planning.

Members of the safety committee have experience and interest in health and safety, and can help you design and improve training. They may be able to help you determine what training is needed and who needs each class. Some members have the knowledge and ability to be instructors, which saves money and enhances the credibility of the training. Best of all, involving the committee in planning, execution, and evaluation of training vastly improves the perception (and the reality) that you are listening to the workers you train.

Safety committees often are looking for meeting topics and speakers. Offer to help them by speaking at meetings, and recruiting other safety professionals you know. You probably have access to health and safety materials they are not aware of, and can design group activities for meetings just as you do for classes. Once committee members are impressed by what you do at their meetings, they will spread the word about your abilities as a trainer.

Two cautions are in order here. Avoid asking the committee to put their stamp of approval on anything for which you have not provided all the relevant information. Since safety is a mandatory subject for collective bargaining, be sure nothing you ask the committee to support circumvents the bargaining process.

3. First line supervisors

Supervisors can make or break your safety training program. If they disagree with what is taught, or with the need for training, they can actively subvert the transfer of training to action. If they are merely neutral about the training, having not been part of the planning, they can passively subvert workers' actions just by their lack of reinforcement of safety changes.

Supervisors are aware of safety problems that need to be addressed. If you are part of their solutions to problems and share or give credit to them for the solutions, they will be more likely to cooperate with your efforts. They may be proud to participate in planning efforts for training if approached with proper deference to their role.

Include supervisors from beginning to end in the training process. If they are part of the program, they will be there to see that changes are made at the action level and will continually reinforce the desired behaviors.

4. Workers

Listening to workers sometimes means you go directly to them and ask them what they need and how you can provide it effectively. In a small workplace, one without a union, or where there is no safety committee, try to involve several workers from each department and shift. Not only can they help you make the training better, they can help you advertise the quality and benefits of the training.

The number one complaint of workers about training designed by safety professionals is that they, the workers, have not been asked for any input. Give workers part ownership of the training program. Worker-initiated training programs go one step further, with complete ownership of the health and safety activities and compliance. You can expect two very important outcomes; better training due to workers' knowledge of the job, and less resistance to training through ownership of the program. They now have a stake in the outcomes of training.

Consider including floor workers as trainers in your program. A number of peer trainer programs are in place in a variety of industries, where thousands of workers, managers, and clerical employees have received excellent training from nonprofessional worker-trainers. Workers have shown they have the ability and the credibility to train other workers, especially when they are prepared through practical, participatory training methods and provided with accurate and interesting training materials they can adapt to their own needs. Think back to what you know about the characteristics of adult learners and the needs of workers: Peer trainers overcome the "expert with a briefcase" hostilities of trainees who want knowledgeable, respectful trainers who understand workers' jobs and concerns (see Chapter 14 for more).

5. Labor unions

Labor unions promote a number of agendas, including the safety and health of their members. Some unions have national safety departments and trainers, and some do not. In any case, the local union can and should be invited to participate in training.

Unions tend to be highly structured organizations, with a hierarchy designed to accomplish their goals. If you are a manager, learn about the union hierarchy, and don't attempt to circumvent it. To involve the union in a safety and health program, start with the president of the local. Let the president delegate responsibility and determine who your liaison will be. If the local president finds the national headquarters has training materials or programs that may be helpful to you, investigate these and use any that fit your learning objectives. Many of these materials have no political context and have been carefully designed to apply to the jobs members do.

Unions are vitally concerned about the safety and health of their members. Local unions can be an important asset in your training program. Involving them in the design of goals and training will serve you well in the long run, especially in enhancing your program's credibility with workers. Credibility leads to acceptance, and acceptance leads to improved health and safety.

Most local unions would like to see more participation in their meetings and activities. Union officers complain that members stay away from meetings in droves, and many don't seem to appreciate what the local does for them. Locals who help plan and implement training that leads to the correction of safety problems identified by their members can publicize these successes; their status in the workplace improves in the eyes of members. When you help them improve status and participation, you give them what they want. In exchange, they support your training program.

VI. Summary

The more groups that have representation in the planning and delivery of worker training, the more likely it is that all the groups the trainer wants to please will be satisfied with the outcomes of training. After all, it is important to satisfy the workers being trained, the supervisors who reinforce the application of the training, the loss control people who collect accident and injury data, and the top management who must show a profit to keep the paychecks (yours included) coming. When you design, develop, and deliver training, include representatives of as many of these groups as you can. The results will be worth your time.

References

1. Hugentobler, M., Robins, T., and Schurman, S., How unions can improve outcomes of joint health and safety training programs. *Labor Studies Journal* 15 (1990): 16–38.
2. Kriesky, J. and Brown, E., The union role in labor-management cooperation: A case study at the Boise Cascade Company's Jackson mill. *Labor Studies Journal* 18 (1993) 17–32.
3. Robins, T., Hugentobler, M., Kaminski, M., and Klitzman, S., A joint labor-management hazard communication training program: a case study in worker health and safety training. In: Colligan, M., ed., *Oc Med: State of the Art Reviews*. Philadelphia: Hanley & Belfus, Inc. 9 (1994) 135–145.

chapter thirteen

Activities for training trainers

The ideal training program for trainers incorporates all the techniques recommended elsewhere in this book, plus activities that increase "face time," the time trainers spend thinking and speaking in front of the class. This chapter describes a number of activities we use in train-the-trainer courses.

I. Goals for trainer training

The desired outcome of training methods/trainer improvement classes is that aspiring or experienced trainers will be more effective trainers after the class than before. Using the guidelines in Chapter 2, let's write some goals for trainer education. Specific learning objectives could be written that lead to the accomplishment of each goal. You may wish to delete, change, or add to the list based on your own experience. Here's what the authors try to accomplish when training trainers.

- Trainers will be able to plan and develop courses and materials that meet the needs of trainees and their employers.
- Trainers will improve their teaching skills and increase their classroom comfort and self-confidence.
- Trainers will be able to adapt to prevailing conditions in human interactions and physical facilities, and be more able to "think on their feet."
- Trainers will develop their training creativity, and gather and retain a body of ideas, activities, aids, and materials to enhance their work.
- Trainers will have fun while they are learning, so they can bring fun into their own teaching.

II. *Working together toward common goals*

As with all your training, the attitude you hope to set in trainer classes is one of cooperative work toward common goals. Budding trainers share many characteristics with all kinds of class participants and, in addition, may be especially uneasy because they anticipate being required to perform. As you spend time together, use your experience and skills to "read" your class and work to generate class unity.

A. *We're all in this together*

If you are training trainers, it is assumed you are an experienced and successful trainer. If these adjectives don't describe you, appoint yourself facilitator and find someone to teach the class who does meet the description. You cannot expect trainers to be interested in what you have read (they could read it for themselves) or what others have told you (where's the proof it works?). They want to know what works for you, the experienced and successful trainer. You want to know what has worked for them in their attempts at training. As with any kind of training, all participants have ideas and experiences to share. It's your job to ensure sharing takes place.

B. *Building class unity*

A fellow trainer had a way of applying his favorite sorting process to every discussion. "There's two kinds of people in the world," he would say, and then use the aphorism to divide everyone into those who agreed with him on a topic and those obviously misguided souls who did not.

There are two kinds of train-the trainer classes — those in which participants know and have worked with each other in an established relationship, and those comprised of strangers. In the first, the leader has a head start in helping trainees build a united group of people who will help each other during the class and perhaps throughout their training careers. In the second, group cohesiveness is built from an introductory level. Here are some unity boosters that have worked for us in both situations.

Every class gives off vibrations that indicate the mood of the dominant, or at least the most verbal, members. Read the mood and, if it suits your goals, work to enhance it. Two recent classes serve as examples.

1. *Class at national union headquarters*

Few of these 41 trainees knew each other; many had come alone to represent their locals. Most of them had never taught; many were fearful about being unequal to the challenge of speaking in front of the group, and several had been sent to the wrong class and thought it was just a basic health and safety class. Almost all were floor workers; of the four managers in the group, all had training experience.

Care was taken to bring everyone together without intimidating anyone, but allowing freedom to those who were ready to generate ideas. We started this group very slowly. The "Find someone who..." icebreaker (Figure 13.1) worked well and started them talking with each other. The leader was enthusiastic and friendly, gave a lot of reinforcement to the feeling that we would all help each other, and carefully avoided calling on anyone specifically for the first day and a half. Small working groups were left together without rearrangement for optimum cohesiveness in training teams. After three days, even those who arrived with no intention to teach said the class would help them when they want to speak at a department safety meeting or a meeting of their local union. Unity developed slowly but positively, especially within the training teams.

2. Company Hazmat team

This was a wild bunch of 21 co-workers with a tremendously outgoing group personality and more creativity than we had ever encountered. They knew each other well, and teased each other constantly. They were also very supportive, and were easily able to give and receive blunt, but not personally unkind, criticism. The leader was able to create challenging problems on the spot that she know the group would handle well, and said she had never taught a class where she laughed so much. In teaching this class, the only concern for the trainer was to keep the decibel level from getting totally out of hand, and to smooth the rough edges from inventive presentations created by training teams who whole-heartedly jumped into devising excellent presentations enhanced by inventive props and methods. This class was unified before the training ever began.

C. Class buzz words

Rather by accident, we learned that our trainer classes like to share buzz words. These phrases, which may come up by design or by divine providence, are repeated over and over and serve as reassuring shortcuts the entire group understands.

1. Tap dancing

Our Train-the-Trainer icebreaker includes the directive "Find someone who can tap dance." All the questions are relevant to training situations, and several lead to the establishment of buzz words. Tap dancing, of course, is what trainers do when materials don't arrive, or the projector quits, or any event disrupts the plan and everyone is waiting for the trainer to fall flat. Many times in trainer classes, when anything goes wrong or there is a question no one can answer, participants advise each other to "just tap dance."

FIND SOMEONE WHO . . .

Has never taught a safety class but wants to.

Has trained more than 200 workers.

Can train in a language other than English.

Has never been embarrassed.

Has taught in a non-workplace setting (school, church, etc.)

Can tell workers about his or her own chemical accident or overexposure.

Has built useful things out of other people's junk.

Can tap dance.

Has obedience trained a Rottweiler.

RULES: Different person for each description. Write down first and last name.

Figure 13.1 All the descriptions on this icebreaker can be related to qualities or experiences of an effective trainer.

2. Kory, sit!

One of the icebreaker instructions is to find someone who has obedience-trained a Rottweiler. Sometimes, the class actually includes such a person. The inclusion of this item on the icebreaker was inspired by an instructor who has trained several dogs, and explains the connection between training workers and training Rottweilers. She describes these efforts.

"Cris, my golden retriever, lived to please. I could call her from across a field, and she would come running immediately, sit in front of me, smile and say, 'Oh please, please, tell me what you want me to do next.' My Rottweiler, on the other hand, looks arrogantly over her shoulder at me and says, 'Maybe I'll come, and maybe I won't. What's in it for me?' I've never trained with food before, but Kory will do anything for a dog treat. Once I found out what she wanted, I could train her." The lesson here: find out what trainees want, and give it to them. They may agree to do what you want to order to get what they want. (Chapters 5 and 12 hold clues to figuring out what trainees want.)

This trainer has been known to put her hand on a student's shoulder and quietly say, "Preston, sit," Be careful when you choose your Preston — this works only with agreeable trainees. Once it has been explained, trainer participants use it on each other when someone talks too long.

3. Bless your heart

This saying is heard frequently in the South, and since the authors have spread it to trainers it is now heard in Wisconsin, Washington, D.C., and upstate New York. It was suggested in Chapter 11 as a response to hardheaded trainees who want to argue, or who just will not stop complaining about their job situation, or with whom one wants to be empathetic while politely shutting off their flow of words. "Bless your heart!" is a complete response in itself. Voice inflection alone conveys the message. Having said it with the proper intonation, and a gentle pat if appropriate, the trainer can, with no further comment, return to her point. Trainer candidates seem to love it.

Those who plan to use the Bless Your Heart response should know it exists at three levels. It would never do to use the wrong Bless Your Heart. Level one expresses real sympathy, as in, "Oh, your puppy dog died. Bless your heart!" Level two is a bit unkind without being obviously so, as in, "Oh my, have you seen Mary since she bleached her hair? Bless her heart." Level three drips with honey, conveying — without having to speak an unkind word — the speaker's opinion of the subject's tacky behavior (tacky is another good Southern word). Just shake your head and say, "That (insert your choice of target), bless (his/her) heart." Southern ladies have the knack for cutting you dead while being polite, as each learned at her mother's knee.

Each buzz word has its own use. "Tap dance" reminds a trainer who finds himself between a rock and a hard place that he has the ability to get out of this uncomfortable situation if he will just breathe deeply and remember what he has learned. "Kory, sit!" reminds us that training it a trade-off between trainer and trainee, canine or human. "Bless your heart" is a useful response to a trainee who is tying up your class with words and worries you do not want to address. It permits a sympathetic response that can immediately be followed by going on with the lesson.

III. Use the method to teach the method

A verbal group critique of a training methods class responded to the instructor's request for last-day ideas on how to improve the course. "For a course that promotes participatory training methods, there's too much lecture on the first day of this course." In the search for participatory ways to teach the advantages and disadvantages of all the different methods, the instructor decided next time to use the method to teach the method. In other words, brainstorm about brainstorming, buzz about buzz groups, and assign problems to small groups about when and how to use problem solving.

A. Assign groups to teach each method

It seemed like a good idea to have a class groups use a method to teach the method, but it didn't work. Methods should be discussed on the first day, but most first-day trainees are not ready to take on the difficulties of leading a brainstorming session, assigning and summarizing buzz group topics, or writing problems to be solved in small groups. If your class is comprised of advanced trainers, by all means ask them to do this, but the people we train, and those for whom this book is written, generally are not ready until later in the week.

B. The instructor uses the method to teach the method

This is good participatory training, with the experienced trainer up front to guide it along. Trainees are actively engaged in thinking, responding, analyzing, and synthesizing. What follows are several examples of using the method to teach the method.

1. Brainstorming about brainstorming

Use a flip chart and traditional brainstorming methods; i.e., ask a question, write down all responses without editorializing or judging, tape sheets onto the wall if more than one page fills (you, or course, have already stuck short pieces of masking tape onto the edges of the chart easel), then group and summarize the points made.

Questions for this activity may include the following.

- What are the advantages of brainstorming as a training method?
- What are the disadvantages?

- What kinds of topics might you use it for?
- What do trainers need to remember as they lead a brainstorming session?

As always, post the sheets and summarize. Lead a brief discussion about what makes a good brainstorming leader.

You may even wish to demonstrate how lack of enthusiasm and good leadership squelches brainstorming. Simply stand there without leading the group and let them watch the quiet roll in and the session fall flat, then ask, "What just happened? How could I have made it work better?"

2. Buzz groups about using buzz groups

Divide the class into groups of three to five trainers, giving each group a flip chart page with a question or topic written at the top. Give them ten minutes to decide on and write four or five responses, after which a reporter will present the group's ideas.

Questions and topics might include the following.

- What makes a good buzz group?
- What kinds of topics make best use of this method?
- What are the advantages to the class of using the buzz group method?
- How can the trainer ensure a good outcome from buzz groups?

3. Discussion about the discussion method

You can do this two ways. In the first, remain at the front of the room and lead a discussion about using discussion as a training method, but focus only on the front left quarter of the class. After a few minutes, open up the conversation by moving around and including the entire class as you ask them, "What am I doing wrong? Why is no one from the back participating?" Then continue the discussion about this training method, making sure to include everyone.

This is very effective, as the class can see your mistakes. Or, if you prefer, just hold a class-wide discussion about the merits and guidelines of discussion as a training method.

4. Role playing to teach role play as a method

Role playing is not an easy training method to use. If you provide a script, participants tend to just read it without improvising and taking on the personas of the roles. If you don't, they may not know what to say. If you have judged your aspiring trainers to be comfortable and uninhibited enough to perform successful roles, try this.

Ask four trainers to volunteer as role players. Give them two minutes to consider their positions (assigned by you) on using role playing in training. Bring all four to the front of the room, and set the stage: They are arguing about what methods to use in an upcoming class. Here are their attitudes.

- Person #1: Playing roles is artificial and dumb. I prefer to just tell people what I want them to know and let them go back to work.
- Person #2: Role play is a valuable training method, especially when it is used to help people learn how to deal with safety attitude problems with supervisors or other workers when they return to work. I can give you some examples of this (then give some).
- Person #3: Most people are too bashful to be dramatic in front of people. I think role play would probably be fun and effective, but nobody I know would do it. Take the people I work with, for example (describe some of them).
- Person #4: Role play is a good method. I just don't know how to do it. Should I assign whoever I want to whatever roles I want them to play? Or should I get volunteers? What if nobody volunteers, or they don't like their roles? How do I decide when it's over, and what should I do then?

Role play is an exciting addition to the class if the players are enthusiastic (Figure 13.2). If you are going to do this activity, decide in advance on your time limit, or determine the point at which you will stop and summarize role play as a training method. On two tracks in your head, keep two things in mind. First, lead the role players toward the accomplishment of your goals for this exercise. Second, keep a running tally of what is going on for your post-activity summary.

5. *Solving the problems associated with the problem-solving method*
Assign each group a problem. Choose problems that describe those you have encountered as a trainer. Here are some the authors have used.

- A trainee group assigned to present a lesson on hearing conservation has chosen a spokesperson. He is up in front of the class using an overhead projector, and is drawing on the transparency, line-by-line, a picture very suggestive of a naked woman. What does the trainer do, if anything?
- Your training group has arrived at the classroom 15 minutes before the start of a large class, only to find that none of the audiovisual equipment you ordered is there. Neither are there any flip charts, or paper for trainees to write on. Only the training manuals, and the aids you brought, are available. Now what?
- Your training partner has not shown up. You have no idea where she is, or what may be wrong. All you know is that you are not knowledgeable about, nor do you have materials to teach, her half of this course. The class participants have gathered from all over the state, and are paying large fees for this course. (You could change the last sentence to fit your group.)
- Your class of 15 people has been totally unresponsive, no matter how hard you work to include them in question-and-answer periods, dis-

Figure 13.2 Role playing adds interest to the class, and reinforces lessons and actions.

cussions, and brainstorming. You were so anxious to get to the lunch break that you called it early, at 11:00. You have an hour to figure out how to handle the afternoon session in this day-long class.

As groups report back, don't simply accept their solutions. Ask "What if?" questions to stimulate them to think further. Create additional problems that extend their original problems to help them continue to look at alternatives. Play devil's advocate, even if you agree with their solutions.

IV. Group activities for trainers

A. Activities that teach planning and development

Without practice in planning and developing lessons, trainers tend to ride off in all directions at once, or in one direction without giving enough thought to whether the ride leads to their goal (if they have one.) Here are some activities that reinforce important steps in planning and development.

1. Writing learning objectives

Include in your training manual the training requirements of the Hazard Communication Standard (Figure 13.3). Plan to divide the class into five groups, and make a large poster of six of the training requirements. We use

OSHA HAZARD COMMUNICATION STANDARD
29 CFR 1910.1200
TRAINING REQUIREMENTS

The following is quoted directly from the standard.

(2) *Information.* Employees shall be informed of:

(i)   The requirements of this section;

(ii) Any operations in their work area where hazardous chemicals are present; and,

(iii) The location and availability of the written hazard communication program, including the required list(s) of hazardous chemicals, and material safety data sheets required by this section.

(3) *Training.* Employee training shall include at least:

(i)   Methods and observations that may be used to detect the presence or release of a hazardous chemical in the work area (such as monitoring conducted by the employer, continuous monitoring devices, visual appearance or odor of hazardous chemicals when being released, etc.);

(ii) The physical and health hazards of the chemicals in the work area;

(iii) The measures employees can take to protect themselves from these hazards, including specific procedures the employer has implemented to protect employees from exposure to hazardous chemicals, such as appropriate work practices, emergency procedures, and personal protective equipment to be used; and,

(iv) The details of the hazard communication program developed by the employer, including an explanation of the labeling system and the material safety data sheet, and how employees can obtain and use the appropriate hazard information.

Figure 13.3 The HazCom standard has very clear requirements for training.

a machine that turns 8.5 × 11 master pages into flip chart-sized posters, but you can carefully print them onto flip charge pages if necessary. Tape the six posters high up on a wall visible to the entire class; under each, tape another poster on which you have written the overall goal for training to comply with that requirement. Table 13.1 shows the text included on each of the posters.

Print up three to five flip chart pages on which you have written possible learning objectives that may lead to the attainment of the first goal. Tape them under the first goal or, if you are running out of wall space, simply leave them on the flip chart. Some of the learning objectives should be good ones (see Chapter 2 for guidelines in writing these); some should look good but not match the goal; and some should be badly written, perhaps

Table 13.1 Contents of the Posters Used in Writing Goals and Objectives for HazCom Training

Section	Required Training Topic	Goal
1910.1200 (h)(3)(i)	Methods and observations to detect the presence or release of hazardous chemicals in the workplace	Workers need to know how to detect the presence or release of hazardous chemicals in their workplace
1910.1200 (h)(3)(ii)	Physical and health hazards of the chemicals in the workplace	Workers should know and understand the physical and health hazards of the chemicals in their work areas
1910.1200 (h)(3)(iii)	Measures employees can take to protect themselves from these hazards	Workers should know how to protect themselves from the hazards of chemicals in their work areas
1910.1200 (h)(3)(iv)	Explanation of the labeling system	Workers must understand the labeling system for hazardous chemicals in their work areas
1910.1200 (h)(3)(iv)	Explanation of the material safety data sheets	Workers must be able to read and understand material safety data sheets (MSDS)

describing learning achievements that are not measurable, or not written in terms of "the learner will." Examples are shown in Table 13.2.

Brainstorm the value of each learning objective with the entire class. Some you will toss out, some can be fixed (do this on the chart with a marker), and some are good as written. Spend enough time with this that each class participant understands the reasoning for each action.

Break the class into five groups. Assign each group one of the HazCom goals for which you did not write learning objectives. They are to write four to five objectives, using the guidelines in Chapter 2. Tell them they do not have to figure out how to completely attain the goal; additional objectives may be needed to actually accomplish the goal. They are to write only four or five. They should write large and legible objectives on flip chart paper. When everyone is finished, ask the group to critique each set of objectives as they did the example set.

2. Rewriting trainee-friendly materials

Choose excerpts from instructional or training materials that seem to you to be overwritten, or that might be made more accessible to a low to medium-literacy level training group. Using the guidelines in Chapter 4, let trainees guide you in rewriting one example you have copied on an overhead transparency. Write the improved version on the flip chart, revising until the class is satisfied with it. Divide into groups of three, and give each group a different paragraph to rewrite using the same guidelines. Tell each group

Table 13.2 Examples of Learning Objectives to Achieve the Goal of Detecting
Hazardous Chemicals

	Learning Objective The Learner Will Be Able to:	Leads to Achievement of Goal?	Measurable?
1.	Indicate on a drawing of the plant where the hazardous chemicals are stored	Yes. If spill occurs, will help to know if it is hazardous by knowing if hazardous chemicals are in the area.	Yes. Ask trainees to point out or name hazardous chemicals on the drawing.
2.	Learn about air monitoring	Yes. Air monitoring is an important detection method.	No. Can't know what trainees "learn." Too general; needs an action verb, as #1.
3.	Describe the appearance and odor of hazardous chemicals in the work area	Yes. Seeing or smelling a release is information leading to awareness of release.	Yes. Ask them to describe these properties.
4.	State the purpose and use of continuous air monitoring devices used in the area	Yes. Area monitoring devices give early warning of spill. Much better than #2.	Yes. Have them state purpose and use.
5.	Know how to clean up chemical spills in the area	No. Has nothing to do with detecting presence or release	No. Change "know" to "demonstrate" and it is measurable.
6.	Know where the MSDS are	Maybe. Being able to look at a MSDS could provide a clue regarding the identification of chemical, i.d. as hazardous.	No. Change "know" to "state" or "demonstrate"
7.	Know which respirator to put on if the chemical spills	No. Not related to detection of spill.	No. Replace "know" with action verb.

what information the users are assumed to have as a prerequisite so they don't feel they have to explain things the learner already would know. There are three possible follow-up activities for this exercise.

   a. Class critique.   Ask the class to critique the revised paragraphs, or have groups exchange paragraphs and critique each other.

   b. Illustrations.   Return the paragraphs and the critiques to the original groups, and ask each group to draw an illustration to help the learner understand the material better.

   c. Scoring. Using the scoring guidelines in Chapter 4, have each group score their written and illustrated material, then revise it to improve their score.

B. Activities that teach self-evaluation

Effective trainers understand their own personalities and abilities as they relate to training. If they are to work in teams, an arrangement that usually enhances training, they should be able to share their knowledge with each other. There are several ways to do this.

1. The self-evaluation sheet

Figure 13.4 pictures a form trainers can use to consider and write down their best skills, talents, and personality characteristics that relate to training. The information can be used to help trainers select methods and materials, and to get to know the abilities of training team members. The items are more fully described in Chapter 5. Special talents may include all kinds of things — anything a trainer could use in the classroom to emphasize a point. He might write and sing a safety song, juggle items that represent the maximum number of distractions a fork lift driver can handle without wrecking, or perform a magic trick that illustrates a lecture topic. A talent for stand-up comedy is useful to a trainer. A really creative trainer can probably stand on his head, or keep a hacky sack in the air, and turn his talent into a teaching enhancement.

2. Videotaping

Some trainer courses routinely videotape student trainers and critique their presentations. If your trainers want to do this, by all means make it available to them. Videotaping makes some participants very uncomfortable, especially when someone other than themselves will critique the outcome. For these, you may wish to skip the idea, or simply give them the tape for their private viewing.

3. Trainer-generated evaluation sheets

If your trainers believe they would benefit from receiving input from their fellow members, ask them to design a sheet to be used when they teach. They include on the sheet whatever items they want the class to evaluate, and may ask for comments or just request yes/no answers. Suggest they include a question asking if they have any irritating mannerisms. Trainers with some of the commonly seen mannerisms may be completely unaware of them. One trainer added the pointless ending "… 'n stuff" to at least one fourth of his sentences. Class participants became distracted from his message and began to keep score. The same can happen with a "you know" habit.

TRAINER'S SKILLS AND TALENTS

Name_____

My areas of expertise: Ability to take control of group
                                    activities:

_____   _____

_____   _____

_____   _____

Material I know well: Ability to keep my head and
                                    summarize if several people talk
_____   at once:

_____
                                    _____
_____
                                    _____
_____
                                    _____

Comfort level in front   My special talents:
of people:

_____   _____

_____   _____

_____   _____

_____   _____

Other things about myself that may relate to training:

_____

_____

_____

_____

Figure 13.4 A self-evaluation sheet helps trainers and their teams assess strengths and talents.

C. Activities to improve instruction and communication

Beginning trainers are so focused on what they are saying and doing they forget to notice whether they are understood. When classroom groups are unable to do the assigned chore, it is very likely the trainer did not give good instructions, is not providing helpful feedback, or has not done whatever it takes to ensure good communication. Here are some activities that point out deficiencies in instructions, feedback, and communication.

1. Paper folding

This activity works better if you demonstrate the first step with a volunteer before others try it. Don't demonstrate the entire exercise, or they will catch on (and you don't want them to). Teams of two stand back-to-back, and are cautioned not to look over their shoulders at each other during the activity. Person one holds a piece of letter-size orange paper, and person two a blue piece (or any two colors that are different). The orange person gives instructions to the blue person about folding the paper. He may repeat the instruction, but not explain it. The blue person may not ask any questions. The orange person folds his paper as he instructs the blue person to do. Neither can see the other's paper. Five separate instructions are given. At the end, orange and blue turn around and compare their folded papers. If they match (which they almost never do), orange has given excellent instructions. If not, his instructions were not clear. Point made.

Have enough paper to repeat this. Trainees like it so much they will want to try again with blue giving the instructions. They usually do better the second-time around.

2. Peanut butter sandwich

Props required are a jar of peanut butter, a jar of jelly, a table knife, a loaf of bread, and a few paper plates. Divide the class into groups of three to six people. Each group receives the following assignment. Write a set of instructions for making a peanut butter and jelly sandwich. You have ten minutes. At the end of ten minutes, you will pass the instructions to another group. One member of the receiving group will read your instructions verbatim, with no elaboration, to a second member of her group, who will follow them exactly. When the sandwich is made, one or more members of the group that wrote the instructions must eat it.

3. Ball toss

You will need six Nerf balls, four prepared flip chart pages, a chair, and a box or trash can for the ball toss. On the flip chart pages, you have printed (or reproduced, using your poster machine) four styles of feedback. The sheets read as follows.

PASSIVE FEEDBACK
Remain relatively silent and passive throughout,
giving little verbal or nonverbal feedback

TECHNICAL FEEDBACK
Provide only specific "technical" feedback and suggestions
to help the thrower complete the task successfully

CRITICAL FEEDBACK
Provide nonproductive criticism any time
the thrower does not succeed

SUPPORTIVE FEEDBACK
Provide as much support, encouragement, instruction
and positive direction as you can

Divide the class into at least four groups — six or eight will reinforce the message even more. The first group selects a representative who sits in a chair facing away from a box or trash can placed six feet behind the chair. The thrower attempts to toss all six balls, one at a time, into the target without being able to see it (although he may look at it once before sitting down). Place the flip chart where the class can see it, but the thrower cannot. The class provides feedback in the style on the flip chart page selected by the instructor.

This is another activity the class will want to repeat. The summary for this activity is obvious, and should include comments from the throwers about how the different styles made them feel, and whether the feedback helped them improve their performance.

4. Dilemmasaurus

The dilemmasaurus is a creature made from Tinker Toys. You could probably build it from Legos also, although we have never tried it. The dilemmasaurus teaches people how to communicate between and among themselves, and is especially useful for training people to relay important information.

Divide the class into at least two groups. Each group has a set of toys and an extra set is required for the original model. Before class, the instructor builds a complex dinosaur-like animal and places it outside the classroom or in another room. (No one should see it before the exercise.) Group members act as communicators, who go out and view the dilemmasaurus one at a time. The communicators are not allowed to make notes or drawings, but must come back into the room and instruct the groups as they build a matching animal. The communicator is not allowed to point, or to touch any pieces. One group member serves as the builder, and does not see the original. You can end the exercise one of two ways: have a time limit, and name the group that comes the closest in the allotted time the winner; or award the prize to the first group to successfully complete the dilemmasaurus. The time limit works best, since successful completion may take a very long time.

If the trainers will be training Hazmat responders, or other groups required to communicate by radio, use radios in this exercise and require the communicators to remain outside the room.

D. Activities that increase "face time" and quick thinking

All the trainers we have ever taught have said the same thing: the more time one spends in front of a group, the easier it gets. Although you will not want to "threaten" beginning trainers by announcing your intention of putting them up front, you will still do it in natural, non-threatening ways. and they will thank you in the end (if not while they are doing it).

Here are some activities that increase face time. Start with getting them to talk in front of people, and gradually steer them into activities that require them to think on their feet.

1. Report back

Whenever a group completes a problem-solving exercise or buzz group, one member reports on the group's efforts. Without instructor guidance, the report-back becomes a list reading. Ask three things of your groups when reporting back.

a. Use a new reporter. Each time a report back is requested, ask that someone new do it. Although groups will not remain static, as the instructor will mix them differently for new group exercises, ask that they always select someone who has not yet reported back for any group.

b. Ask for a summary or interpretation. Ask reporters not to just sit or stand in place and read the group's list of conclusions. For the first one or two, have them stand in place; after that, ask them to come to the front of the room. Request that they summarize, or put their own spin on the report, rather than just read the list. They should look at and speak to the class, using notes only as an occasional reference. Be sure to make this request at the beginning of the group exercise, or when you announce time is almost up. The reporter needs time to gather his thoughts.

c. Ask for part of the report. It has been mentioned earlier that multiple report-backs quickly become redundant and boring, especially if all groups used he same worksheet. Ask each reporter to tell about one portion of the exercise. The instructor then summarizes them into a complete whole at the end of reporting. Explain to the class why you do this, and suggest they use the same method when they train.

2. Grab bag

Fill a pillowcase or paper bag with a variety of small items. They may include a deck of cards, a small set of Legos or building blocks, a yo-yo, a sack of Nerf balls, a group of colorful bar napkins and/or coasters, a Slinky toy, an unusual hat, or any item that looks like it might be interesting. A representative of each group of three to four aspiring trainers reaches into the back and blindly chooses an item or set of items. The group has ten minutes to devise a presentation for which the item serves as a visual aid. Don't do this on the first day if your trainers are beginners; wait until they feel a little more confident. If the class is small, have each trainer do this individually, or with just one partner.

3. Microphones

Play "Drop the Hatchet," a childhood story-telling game. Here's how it works. Holding a non-working microphone, one member of the class starts a story, or, in this case, a training lesson. After setting up the scenario, she says "Drop the Hatchet," or perhaps in this version, "Pass the Microphone" would be better. She passes the microphone to any member of the class she chooses, who has to take up the lesson and continue. Speakers may go in any direction they wish with the lesson. Creativity is encouraged. Microphone holders should stand up and speak loud enough to be heard by everyone in the class, without having to come to the front, as rapid continuation of the lesson is important.

Use this exercise with experienced trainers, or after several days of groundwork with beginning trainers, to give them practice in quick thinking and immediate response. Lacking a microphone, designate a substitute object.

4. Egg toss

Buy some plastic Easter eggs that open, or ask friends to save for you some of those panty hose containers that look like eggs. Inside each egg, place one small piece of candy (or weight and reward), and a piece of paper on which is written a topic or assignment. Toss the eggs one by one into the class. The people who catch them have two minutes to make up a two-minute presentation on that topic. Time them, and keep them inside one to three minutes.

For an alternative egg toss, each inclusion states a position for a role play. For example, "You are an hourly worker and a member of the safety committee. You are getting tired of the run-around you keep getting about a continually wet area of the plant floor. Several people have slipped there, but so far, non have been badly hurt." Other assignments in this role play might be the safety manager, who is overworked and over budget; the shop steward for the local union; and a worker who has slipped there twice and is pregnant.

5. *Cheers and songs*

Give each group of three to four people 15 minutes to write a class cheer, class slogan, or class song. They will teach it to the class, who will practice it under the writers' direction until they get it right. If you want to encourage active (and possibly loud) participation, steal a trick from cheerleading camps attended by high school cheerleaders and award a "spirit stick" to the most enthusiastic group. The spirit stick can be anything you want it to be. Pass it around during the several days of class to the most spirited group, or the group with the best ideas, or the group with the quickest finish or highest score in any competitive training activity.

Cheers and songs also promote class unity. Nothing is more heartwarming than to wave goodbye to a class of trainers as they tromp down the stairs singing new words to a tune usually rendered by a popular purple dinosaur: "I love work, work loves me, we will all work safe-uh-ly." It's enough to bring a tear to the eye of the most hardened trainer!

Caution: not every class does this well. Don't attempt it if your group is very dignified, but give a quiet group the benefit of the doubt. Perhaps inside each there lurks a showman.

6. *Training teams*

By the end of the class, beginners and experienced trainers are ready to train. Divide them into groups, being careful to use what you have learned about them to ensure that each group contains a good mix of idea people, articulate people, and people with access to gadgets and equipment at the training site. Give them plenty of time to plan and prepare a 30-minute team training session. If your class includes people who still are very intimidated by public presentations, tell them not all members must teach as long as all members contribute to the planning. Provide them with a lesson plan form (Figure 13.5) for team training, which they will turn in to you just before they start the session. If you have time, have the class agree on some points they would like other class members to use to critique their session. These points should be very general; for example, "Did the methods chosen suit the training topic?" and "Do you have ideas the group could use to improve their session?"

When assigning team training, we set only two guidelines: use a participatory method (no lecture); and stay within five minutes of the allotted time. If a team goes too long, we stop them. If they finish too soon, we simply say "Your next trainer is not here yet; he starts at 11:45 (or whenever), then sit back down to let them figure out what to do. We sit in the back as part of the class and let them handle inattentiveness and other problems that occur, only interfering if things begin to get out of hand and the team is not responding. Then we might walk forward and say, "How might you handle this problem? The people in that corner seem to be having their own conservation."

LESSON PLAN: TEAM TEACHING

The team teaching practice lesson will be limited to 20 minutes, including questions and discussion. Use any or all of the methods you have learned. Include everyone in your group in the planning, design, and development, and in the teaching if possible.

The members of this class are to pretend to be whoever you want them to be. They are workers in your facility. You decide on what their jobs are and how much training they have had.

Topic
Name the topic you will teach.

Goal of the lesson
Write the overall goal of your lesson.

Learning objectives
Upon completion of this lesson, the learner will be able to:

Prerequisites
The lesson assumes the learner already knows:

Figure 13.5 The lesson plan is used for team training practice.

State the method or methods you have selected to use.

Reasons for selecting method(s)

Explain why you chose to use this method, basing your selection on the topic, the class members, and the trainers in your group.

Materials needed

List the materials you will use.

List the materials you would add if they were available (in other words, what would you use if you were at your own place instead of at this class?

Critique

The class will help you decide if what you did worked, if they got the point you intended, and how you might make this lesson more effective the next time you do it. Is there anything else you want by way of critique?

Figure 13.5 (Continued).

E. *Activities that teach evaluation*

Evaluation is not usually foremost in the minds of beginning trainers. They are more concerned with immediate problems, like what to say and whether they can survive the class. After years of being trainees with no opportunity for input or, at the most, "smile sheets," they may have little idea how to accomplish any kind of meaningful evaluation of their training efforts.

1. *Evaluating examples*

Provide a worksheet (Figure 13.6) and examples of evaluation forms from various training organizations. Give each group 20 minutes to fill out the worksheet. This exercise works much better than simply telling them about effective evaluation forms. Of course, as a proponent of participatory training, you already knew this.

2. *Designing an evaluation for the train-the-trainer course*

There are two ways to do this. Both start with small groups who are asked to design an evaluation form for the course. Tell them what you, the trainer, want to do with this form. Before class convenes on the final day, use their forms in one of two ways.

a. Combine group efforts. Overnight, combine the ideas of all the groups into one cohesive evaluation form. Give everyone a copy, and ask the class to use it to evaluate the Train-the-Trainer course. Chapter 10 includes an example.

b. Use all the forms. You may find the student-designed forms are so different they cannot logically be combined. Copy them all, stapled together with identifying numbers or letters on each. Pass them out, and ask each trainee to choose one and complete it. When they have done so, and while they still have the forms to look at, ask them to tell you why they chose the form they did. It seems to be the case with our trainer-trainees that they choose the shortest, simplest form.

We learned something from their choices. Trainees tell us that, if the evaluation is long and complex, they tend to just circle whatever their pencil hits and don't provide the requested written comments. By the time the evaluation is handed out, they just want to go home. This information has led us to provide our evaluation forms earlier so that trainees have time to complete them in a thoughtful manner, and we suggest they do the same when they train.

3. *Write a follow-up evaluation form*

Trainees have been exposed to the ideas of process measures and outcome measures in the section on evaluating training (Chapter 10). Now we ask them to design a form that will be mailed to trainees at some interval following training. They are asked to be very specific about what they want

WORKSHEET
EVALUATING THE EVALUATION FORMS

Evaluation form #_____

Please answer the questions and make comments.

1. Is the evaluation form easy to use?

   Are the instructions easy to follow?

   Are the questions clear (not ambiguous)?

   Are the questions stated simply?

   How about the literacy level?

2. Does the form ask for opinions or information you think students would be qualified to give?

3. Could you see the reasons for all the questions?

4. What do you think is the overall goal of this evaluation form? In other words, what do the trainers want to do with the results?

5. Would you use this form as is; if not, how would you modify it?

Figure 13.6 Trainers evaluate evaluation forms using this worksheet.

to know, not just whether the workplace is safer but what aspects of workplace safety they want to measure. They also describe how they will increase mail-back, since we know that over 50% of recipients will not return the form.

V. Summary

Use participatory methods when training trainers. Practice what you preach. If you take to heart the premise of this book, that all learners have something to contribute, that people learn best when they are comfortable and having fun, and that participatory methods enhance learning, design your train-the-trainer classes around these guidelines.

As a trainer of trainers, your ideas are exciting and probably will enhance the training you do. Try them. What's the worst that can happen? Perhaps they waste 30 minutes of class time, or someone thinks you were silly. So what? At least you let your creative juices flow — you don't have to do that particular activity again if you think it was a waste of time.

Each time we train trainers, we learn from them. Some of the activities in this chapter were first seen when student training teams presented them in classes. Each time we try a new activity we have read or heard about, we learn from the attempt; even if we decide its contribution was marginal and we won't do it again, the analysis of why it didn't work is valuable. We have learned to say, "I'm not sure that worked. Do you think it helped you?" In a classroom where the mood is relaxed and unity has been established, honest feedback will follow. Help your trainers establish the habit of watching anyone they observe speaking or teaching to learn what makes them effective. During a training methods class, stop and ask "Why did I do that? What was the result?" or "How could I handle this?" or "What made that work?"

One more time: We can train workers and trainers successfully while everyone, including all the trainers, has fun. We do this by using the same methods to train trainers that we want the trainers to use when they train.

chapter fourteen

The case for peer training

There are far more workers in facilities where safety and health hazards exist that there are trainers who can reach them. A variety of training options have been used to solve this problem; the most effective, in the opinion of the authors, is the use of peer trainers.

Peer trainers are accepted by other workers because they can satisfy the needs expressed by workers who are asked to describe accessible training: knowledge of the job; acknowledgment of equality with class members; understanding of the stresses and constraints of working under marginally safe conditions; frustrations with supervisors and management; and the effectiveness of training with immediate application to job skills and life experiences. Two case studies illustrate how effective peer safety and health training can be.

I. Henry and the Hazwopers

In 1987, the National Institute of Environmental Health Sciences (NIEHS) awarded training grants funded by Congress through the Superfund Reauthorization and Amendment Act to train workers covered under the Hazardous Waste Operations and Emergency Response Standard, 29 CFR 1910.120, commonly known as "Hazwoper." The standard mandates training for workers who remediate sites contaminated with hazardous wastes; workers who handle hazardous waste at facilities with Resource Conservation and Recovery Act permits to treat, store, and dispose of wastes; and workers engaged in chemical emergency response. Grants were awarded to nonprofit training organizations such as universities and labor unions, and the progress of trainees carefully monitored by the NIEHS, the Environmental Protection Agency, and Congress. These grant-funded programs produced a large pool of well-trained peer trainers in a variety of agencies and industries.

Four employees of a paper company attended the first regional class organized by a large international union and taught by safety professionals from a university labor education center. The class covered First Responder

Awareness Level training as described in the Hazwoper Standard, with additional training and practice in training workers. On their return to the plant, three of the employees, a pipe fitter, a maintenance oiler, and a quality assurance tester, approached management with a proposal to initiate a training program. They recruited other interested workers, sent them to subsequent classes, and organized a training team to give hands-on instruction in recognizing hazardous substances and responding rapidly to their release. The initial classes were well received by hourly workers and managers who attended, and students quickly dubbed the training team "Henry and the Hazwopers."

Henry, the initiating spark and team leader, says "The company basically said, 'We've got enough trust in you, go with it.' We're proud of the fact that we got everyone in the mill up to an emergency response awareness level in the program's first year." The Hazwopers trained every employee, including managers, sales and technical personnel, and clericals, in multiple training sessions over five weeks, using materials they adapted from the university training manual. Together with the company's training and development manager, the team produced a video and wrote a manual in extensive preparatory work before the sessions. Attendance was mandatory and ranged from 40 participants to 90. Each session saw intense training where every employee was given exercises and various tasks to complete. "It was hands-on and interactive," says Henry. "We kept the classes busy through participation with questionnaires, problem-solving exercises, and using references like the *DOT Emergency Response Guidebook.*"

The company safety manager says, "They created a response to every imaginable kind of incident here. They taught an awareness of not just how to respond, but an awareness that convinced workers that this is part of their jobs." The divisional industrial relations manager called attention to the mill's long string of 1.7 million hours without a lost-time accident, the longest run in ten years, and credited the Hazwopers. "They took on quite a load," he said. "Not just in getting educated on this subject and then presenting it as professionally as they did, but proving there's lots more talent in this mill than just performing one's normal job." The human resources manager placed rank-and-file skills into practical terms. He calculated that if the company had to contract out such training "it would cost a bundle," estimating a professional trainer would charge over $1000 a worker for the depth and scope of instruction mill employees received. "Economically, it's obvious it was quite an advantage to the company."

The mill manager, who was totally supportive of the peer training team, says "It's easy for management to dream up approaches for such training, but the problem becomes how to get it institutionalized. Any employee-driven programs tend to come across with credibility, thus acceptance and effectiveness." Says the supervisory fire marshal who heads up the Hazmat team, "The enthusiasm and energy of these folks is what made this program take off. There's no question as to its value now."

Training team members learned a great deal. Says one, "It was challenging. I learned a lot I didn't realize in a number of years around this mill. This program gave me a great number of tools I needed to effectively share with other workers." Another, explains "What we had to offer was exactly what the company was looking for. It was employee involvement to the fullest."

During the third year of training, an Employee Safety Council (ESC) was formed with three divisions. The Safety Awareness and Communications branch coordinates all safety promotional aspects, reviews all safety statistics and recommends improvement methods, and maintains an employee safety feedback system. Audit and Observations conducts and assesses a mock safety audit based on corporate audit criteria, performs monthly PPE audits, records noncompliance, and communicates results of audits. The Training and Procedures Development division coordinates safety and health training for ESC members, summer employees, and new employees, and standardizes safety and health procedures. They also maintain records of training profiles.

Throughout the mill, departmental trainers conduct classes using kits designed by the original team. After testing the usability and application of the kits and receiving feedback, Henry and the Hazwopers designed training kits for 70 different safety and health procedures and a variety of department-specific training applications. Each kit outlines the key critical elements within each procedure. The trainers love the kits, because they make training straightforward and easy. One of the departmental trainers who uses the kits said, "The materials are very good; the overheads are well put together and I am pleased this was made available to us."

Henry, who initiated the program and led it from the beginning, is thoroughly pleased. Employees are involved throughout the entire safety structure and feel a strong sense of ownership of the safety program. Hourly employees developed all the training kits, and conduct all safety training. The facility has 100% peer-led training. The incidence and severity of accidents and injuries has declined measurably and, according to both management and the peer training team, prove that record production and safety can co-exist.

II. Mike's Hazmat team

Mike, a shift supervisor in a pulp mill, became aware of a large void in the Hazmat response capabilities of his company. A few operators in the mill were marginally trained for chemical responses in the pulp mill, but a large void existed in the remainder of the mill. As a long time hourly worker, Mike had never liked being forced into a new role, especially one with the potential for serious chemical hazards, and he wanted to form an all-volunteer Hazmat brigade. At this organized mill, he knew he had to bring the union on board for a volunteer team to succeed.

In a company newsletter included in every weekly paycheck, Mike announced eight meetings he set up at different times of the day and night where he explained his plan. About 30 people volunteered to train and become part of the new hazardous materials response team. They were told up front this would be a team effort, and team members would be involved in all decisions involving planning, equipment, standard operational procedures (SOPs), and all facets of response. Mike confides writing the plan would have been easier if he had written it all himself, but states "it probably wouldn't have been as good," and certainly it would not have served his purpose as well. Team ownership was a primary goal.

Hazmat team members completed the required OSHA training, and within six months were ready to respond to chemical incidents. They selected all personal protective equipment, wrote team SOPs, and made all decisions, large and small, regarding such things as how to manage the chemical-protective suit use, where to keep the equipment, and who was responsible for maintenance and inspection. All team members were Incident Command trained and all are qualified to lead the team in an emergency; although not every member wants to serve as incident commander, all "talk the same language" and understand each role they may be asked to play.

Little by little, the team began to run itself. Mike carefully increased individuals' responsibilities and watched for leaders to emerge. Small teams within the group volunteered to research important topics and share them with the team, leading them to take stronger roles and respect the abilities of other team members. When presentations and training were scheduled, team members helped Mike prepare. Soon they were helping plan and implement drills, and writing scenarios for practice — everything short of being the up-front person who introduced the drill. They looked at themselves as having pieces of knowledge, but not yet as trainers.

Fortuitously, an opportunity arose to participate in a two-day train-the-trainer session provided by their union and taught by university-based safety trainers. The timing was perfect for Mike's team to have their first exposure to guidelines and practice for becoming trainers. Their successes in the class, and their creative team training practice sessions, fired them up. They convinced the company to contract with an instructor to come to their facility and provide a three-day training methods course rolled into a week-long meeting to plan their year of Hazmat team training, which would be planned and accomplished by the members themselves. The mill manager showed his support by dropping by for lunch with the class, and participating in one of the team-building exercises.

Now the team took charge. They formed small teams to lead monthly skills training sessions, and took over planning, communication, and training room duties. One group acquired responsibility for the team's equipment, including ordering, cleaning, and pressure testing totally encapsulating suits. Another organized flammable liquids response training with local firefighters and convinced the company to hire the town fire chief as a consultant. Others studied and presented incident command refresher training, taught the use of new air monitoring equipment, and designed a full-blown field spill incident for company-wide participation.

The team now has its own training office, computer, telephone, and rotational schedule for manning the office. To this point, training has been limited to the other team members but plans are in place to do the company's Hazard Communication training. Beyond that, the team is preparing to become more involved in all the regular safety training for the mill. This group of hourly workers has gradually become a confident, effective team of outstanding peer trainers.

III. Cooperation in peer training

Neither of the case studies reported here could have taken place without a strong cooperative relationship between labor and management. Whether initiated by a union, a group of workers, or company management, research has shown that several factors must be in place for peer training to succeed.

True input and ownership are key to successful peer training. No matter who introduces the project, if workers are not included in the program design from the very beginning, the effort is unlikely to produce the desired results. The strongest peer training occurs where workers are in control, with management serving as a resource for financial and technical support. In many of the organizations where workers are trainers, supervisors and managers participate as trainees and reinforce the training both before and after class. Research has shown that when management initiates joint programs, worker acceptance and continued support depend on playing a strong role in planning and implementation, and seeing discernible results.

IV. Summary

A close look at the two case studies presented here has convinced us that workers can become excellent worker trainers. Not only do they demonstrate dedication, knowledge, and administrative abilities, their understanding of the safety problems of individual tasks within their facility far surpasses that of a trainer who comes to training without benefit of experience as a floor worker. An especially good combination pairs peer trainers with experienced trainers who have backgrounds in the health sciences and who can answer technical questions regarding chemistry, toxicology, and industrial hygiene practices.

As Ralph Johnson, director of the Center for Labor Education and Research at the University of Alabama at Birmingham, has said, "Workers have shown they have the ability and the credibility to train other workers." Our experiences with Henry and the Hazwopers, Mike's Hazmat Team, and a number of other peer trainers has certainly shown that to be true.

chapter fifteen

Venturing off the map

We wrote this book for Mark, who needed a training map to follow. Mark is not alone in his situation. Perhaps you will recognize his story.

Mark was hired right out of college to be the environmental manager for a small company with plants in four different corners of a mid-size state. The company makes steel tubing, paints some of it, and has comparatively few hazardous chemicals onsite, but enough to make them a large quantity generator. The largest plant employs around 175 people. Most of them work in the tube mill; of the rest, 30 work on the paint line in four shifts, two handle drums of hazardous waste, and 15 work on the drum press line in two shifts. The drum press is close to large tanks of solvent and waste solvent.

Mark is a victim of "By the way . . ." as are many of the readers of this book. "By the way, Mark, find out if maybe some of our guys need OSHA training, and teach them whatever it is they need to know." Mark asked somebody he went to school with where to find out about OSHA regulations, and started looking for information — in between his other tasks — writing a Spill Prevention and Contingency Plan; endlessly documenting the company's environmental compliance; keeping track of environmental regulations; and hosting inspectors from the state environmental agency. Feeling overwhelmed by the volume of OSHA regulations that imply or require training, he heard there was a good training organization in town and called them for advice. Many phone calls later, and after an unannounced visit to the training director's office, he ceased circling, and together he and the training director produced a minimum training needs table. The plan took into consideration the extensive overlap between DOT, EPA, and OSHA training requirements (see Chapter 2), and proposed two one-day training sessions for 22 employees each, at a total cost of around $6000. Mark was relieved: he finally knew what he had to do. The training organization was pleased: their input of many hours of free advice might show a return.

Mark's supervisor said no way. Where was the money going to come from? How could he possibly free up that many people? You can write the rest of the dialog; you've heard it all before. Mark was told to do the training himself, in his spare time.

It is for Mark that we wrote this book. It is a road atlas, a map with clear directions for getting where you want to be. If Mark manages to squeeze out enough time to go over the training plan in the vendor's proposal, he can at least write some measurable learning objectives and purchase materials that will lead to their accomplishment. If he convinces the company to hire a training manager for the four plants (he's working on it), that person can not only write a set of learning objectives, but also develop teaching materials to implement them. We hope this book will help. But before we go, we should make you aware of a different approach to training offered by a number of fortunate people who thrive in the rarified atmosphere beyond normal earthly training limits.

I. Advice from another planet

Somewhere in this universe there exist training managers who have the time and resources to do all the training they believe will lead to the achievement of their goals, whether they be reduced lost-time injuries or employee buy-ins to a behavior-based safety program. These individuals are relaxed, competent, comfortable in their technical knowledge and communication skills, and willing to engage in moderately risky behavior when face-to-face with large groups of potentially hostile employees. If you are, or will admit to wanting to be, such a trainer, follow us as we project our thinking off the map we've been following for 14 chapters. This chapter will contain no interstate highways, or even blue-line roads, no prearranged destinations or suggested routes. As noted on early sailing maps, off the edges of the known world the map reads "HERE BE DRAGONS."

A number of people in the training field are considered experts by other trainers. These are the people interviewed in business and training publications for responses to questions such as, "What is the best way to train?" and "Where is employee training going in the next century?" Most of them work for large corporations, universities, or their own consulting firms; many have degrees in instructional design or educational psychology; all have years of experience in applying their knowledge to all sorts of training populations. The ideas in this section come from them.

The balance of work shifts in this chapter. Previous chapters have painstakingly directed the design, development, and delivery of training. This chapter provides very little direction, because maps are incompatible with unmapped training. Most trainers have never thought about taking an unplanned route to a destination, much less a route that cannot be planned because the destination is not yet chosen. Trainers who dare to try the ideas described here will have to plot their own way, figure out what to pack, and determine how to know when they arrive. The chapter, therefore, is short.

A. Ideas from the experts

All the quotes in this section were first printed in *TRAINING*, February 1999, and are used with the permission of Lakewood Publications.

"Information is not instruction." — M. David Merrill, professor in the department of instructional technology at Utah State University

"John Dewey lamented in 1916 that learning by doing is the only learning method that works, and that nevertheless the schools kept trying to teach 'by pouring in.' We still think that if you tell someone something, then he knows it. Every time a stand-up trainer gets in front of a group and talks about . . . the right way to do a job, he or she is wasting everyone's time." — Roger Schank, director of Northwestern University's Institute for the Learning Sciences

"Classrooms are generally a waste of time. We humans have done an extraordinarily effective job in nearly destroying the innate desire (to learn). We insist that learning requires transmission — from expert to novice, teacher to student. Even when we know that the root word *educere* means 'to draw out,' we've stayed focused on pouring in." — Margaret Wheatley, principal in Kellner-Rogers & Wheatley, Inc.

"Training achieves greatness when it echoes the voices and concerns of employees . . . and when it is rich with the challenges and opportunities that actually happen in . . . the plant. Great training demands much of participants in thought and action." Allison Rossett, professor of educational technology at San Diego State University

" 'Conducting training' is an input, and often an irrelevant one. 'Enhancing performance and productivity' seems much more accurate, since it's an output." Alan Weiss, president of the Summit Consulting Group

"In many cases the content actually needs to be invented by the learners. We inadvertently may be creating 'disabled' learners when we spoon-feed them instruction, in the form of one-way information dumps, in an effort to achieve homogenous outcomes ('Upon completion of course, everyone will have learned to behave in exactly the same way'). We need to be more like air-traffic controllers for spontaneous learning, instead of playing the role of pilot and tour guide who takes people to a predetermined performance destination." — Diane M. Gayeski, principal with OmniCom Associates and associate professor of organizational communication, learning, and design at Ithaca College

"Behavior change takes weeks and months, requires sustained motivation and effort, and, ideally, on-the-job practice in a supportive environment." — Daniel Goleman, co-chair of the Consortium for Research on Emotional Intelligence in Organizations at Rutgers University's Graduate School of Applied Psychology

"For performance problems involving experienced job incumbents (as opposed to novices), training is a less likely solution than nontraining interventions, like better feedback or work-process reengineering. Implication: a 'training needs assessment' is *not* the first step. A diagnostic front-end analysis needs to examine the causes of the performance problem, not just ask what kinds of training people might need." — Joe Harless, president of the Harless Performance Guild

B. *"How do they do that?"*

A *Doonesbury* cartoon published during the Reagan years pictured the president, who sometimes was caricatured as being out of touch with the real world and certainly not bullish on environmental protection, speaking to an aide as he observed dead fish in a polluted stream. The conversation went something like this.

"How do they do that?"

"Sir?"

"That. How do they get those little Xs on their eyes?"

The same feeling of being in a parallel universe, outside of everyday parameters, may strike trainers as they read the experts' comments on training. How do they do that? Where should one begin? What items do I put into my box to take into the classroom?

Common ideas unify the comments. With apologies for putting interpretive words into the experts' mouths, perhaps they are related to suggestions from previous chapters.

1. *Forget stand-up instruction and one-way teaching*

Instead, make it possible for learners to learn. More specifically, don't lecture. Choose other methods from Chapter 5. The trainer using participatory methods becomes a guide, not a dictator. He is one of a group of people gathered to solve a common problem; he solicits contributions from them to a consensus on defining the problem, then helps to organize a framework for solutions.

This is easy. Any trainer can match the experts on this one. The more experienced trainer is more comfortable in a flexible guiding role than the novice, but the beginner who is brave enough to give it a try soon becomes the comfortable, experienced trainer.

2. *Everyone builds the agenda*

Rather than setting an iron-clad, time-specific agenda for a session, help it evolve with input from everyone. Seek input, both formally and informally, from all the stakeholders in a safety situation early. The key words here are "all" and "early." Seek input with open-ended questions rather than nominal approval of a schedule already written (Chapter 12).

Trainers who travel to teach a prearranged class, and instructors who design open-enrollment courses consisting of workers from different companies, will find it difficult to solicit prior input from participants (although with advance contact or sequential classes this is possible). A prearranged set of topics may be necessary in certain situations; perhaps a middle ground can be attained. List the topics in the anticipated order of presentation, without time designations, with the expectation that participant interest and input will determine the length, and to some extent the direction, of each topic. Keep yourself free to rearrange the order based on interest if doing so does not detract from the overall scheme.

Controlling instructors find this idea difficult to entertain, and overly structured trainers are unable to deviate from the agenda. A safety director once returned an agenda to the authors, requesting that times be written in and each break clearly scheduled. A trainer has been observed to shut down interest-driven questions about a topic listed for later in the course, saying, "We'll cover that after lunch." We regret the reluctance to be flexible demonstrated by both these individuals.

3. Trainees set goals and outcomes

Begin the approach to training by asking trainees questions, and incorporating their answers into a set of goals and desired outcomes for the session. Possible questions might look like this.

- How can we best define the problem that led us all to be here?
- Having defined the problem(s), what do you think can bring us to solutions?
- What do you personally need to know, discuss, or practice to work more safely on that problem?
- How do you think I can help, and who else do we need to involve in our meetings?

Goals and desired outcomes are then posted in the meeting room or work area, and referred to often for assessing progress. Goals and outcomes may change as the session progresses, and new ones set for subsequent meetings. If people in your classes obviously are not engaged in the session, perhaps your guess about what they wanted or needed was wrong. Take a radical step: Ask them. Chapter 12 contains suggestions for soliciting input from trainees, as well as others with a stake in the outcomes of training.

4. Make performance the starting point

Look at on-the-job performance, not regulatory training requirements, as a starting point. Contrary to the statements made to workers in some training programs, there are no OSHA requirements for numbers of hours of necessary training (except for minimum and implied hours for 1910.120). Document training hours, of course, but avoid being driven solely by them.

This suggestion is more easily accomplished by a plant-based safety trainer than by travelling or open-enrollment course instructors, though a consultant who is also a trainer could provide this kind of assessment leading to improved performance.

Make workshops, problems, and hands-on training expressly relevant to the tasks and worksite (Chapters 7 and 8). Use task competency as an evaluation criterion (Chapter 10), not documentation of hours of training. If competency criteria have already been met, spend meeting/training time in activities participants select for improvement.

C. Final thoughts

The challenging world of safety training awaits you. Mark your calendar with a reminder to read this book again every six months or so. The assimilation of new ideas is dependent on the circumstances in which you find yourself each time you read a chapter. Every review will bring information you don't remember reading before, and suggestions you may be ready to try. Everything we try makes us better trainers, even attempts that end in less-than-maximum successes.

Remember: We're all in this together, seeking more effective ways to help workers work more safely. We can have fun along the way, and so can the people we train, or what's the point? We have to go now. Mark is on the phone again.

index

A

Acetic acid, rail yard spill exercise, 160, 164
Action plans
    advantages, 74
    disadvantages, 74
    guidelines, 74
    overview, 72–74
Adult learner
    characteristics, 57
    training method summary, 86–89
Adventure, view of training, 6
Aerial Location of Hazardous Atmospheres
    (ALOHA), 216
Air surveillance exercise, 141–144
ALOHA, see Aerial Location of Hazardous
    Atmospheres
Ammonia, rail yard spill exercise, 160, 165

B

Baby Boomers, features, 92
Ball toss activity, trainer training, 291–292
Benzoyl peroxide, rail yard spill exercise, 160,
    166
Blooms taxonomy of thinking process,
    application to learning objectives,
    14–16
Brainstorming
    advantages, 66
    disadvantages, 66
    guidelines, 66
    overview, 65–66
    trainer training, using method to teach
    method, 282–283
Budget
    payments to trainers, 38
    training, 7
Buzz groups
    advantages, 66
    disadvantages, 67
    guidelines, 67
    overview, 66
    trainer training, using method to teach
    method, 283
Buzzwords, trainer training
    bless your heart, 281

    guidelines, 281–282
    sit, 281
    tap dancing, 279

C

CAMEO, see Computer-Aided Management
    of Emergency Operations
Caring, requirement in trainers, 90
Case studies
    advantages, 67
    disadvantages, 67–68
    guidelines, 68
    overview, 67
CD-ROM, research and reference, 212, 215
Cheers and songs activity, trainer training,
    295
Chemical demonstrations, 118
Chemical hazard, assessment and
    worksheets, 151–156, 190–191
Chemical spill, rail yard scenarios, 160–177
Chlorine cylinder training device
    pressurization, 198–199
    repair, 199–200
Chlorine ton container, training device,
    201–202
Chloroform, rail yard spill exercise, 160, 167
Classroom
    arrangement, 21, 251–252
    equipment, 252
    opinions of experts, 309
    physical comfort guidelines, 252
    physical location, 39
Classroom stations workshop
    container identification exercise, 131–136
    participatory training, 113–115
    personal protective equipment
    workshop, 120–130
Computer
    components, 209–210
    computer-based training
        advantages, 220
        disadvantages, 220
        guidelines, 221
        overview, 220